D0576622

ECONOMIC MODELS OF AGRICULTURAL LAND CONSERVATION AND ENVIRONMENTAL IMPROVEMENT

ECONOMIC MODELS OF AGRICULTURAL LAND CONSERVATION AND ENVIRONMENTAL IMPROVEMENT

Edited by Earl O. Heady and Gary F. Vocke

IOWA STATE UNIVERSITY PRESS / AMES

The late EARL O. HEADY was the Director of the Center for Agricultural and Rural Development from its founding in 1958 until his retirement in 1984.

GARY F. VOCKE is an economist at the Economic Research Service, United States Department of Agriculture.

Frontispiece: Earl O. Heady (courtesy Center for Agricultural and Rural Development, Iowa State University)

Manufactured in the United States of America
⊛This book is printed on acid-free paper.

First edition, 1992

Library of Congress Cataloging-in-Publication Data

Heady, Earl Orel
Economic models of agricultural land conservation and environmental improvement / Earl O. Heady and Gary F. Vocke. – 1st ed.
p. cm.
Includes bibliographical references.
ISBN 0-8138-0523-6 (alk. paper)
1. Soil conservation–Economic aspects–United States–Mathematical models.
2. Agricultural ecology–Economic aspects–United States–Mathematical models. I. Vocke, Gary. II. Title.
HD1765.H43 1992
333.76'16'0973–dc20 90-30228

CONTENTS

SECTION IV.
APPRAISAL OF THE LONG-TERM PRODUCTIVITY OF THE U.S. AGRICULTURAL LAND BASE

SECTION V.
INCORPORATING DEMAND RESPONSE IN ENVIRONMENTAL MODELS OF U.S. AGRICULTURE

CONTRIBUTORS

William G. Boggess in Professor of Food and Resource Economics, University of Florida.

Carl C. Chen is Associate Professor of Operations Research at Indiana University of Pennsylvania.

Burton C. English is Associate Professor in the Department of Agricultural Economics and Rural Sociology, University of Tennessee, Knoxville.

James A. Langley is Assistant to the Deputy Administrator, Program Planning and Development, Agricultural Stabilization and Conservation Service, USDA, Washington, D.C.

Anton Diderik Meister is Reader and Associate Professor in Natural Resource and Environmental Economics, Department of Agricultural Economics and Business, Massey University, New Zealand.

Vishnuprasad Nagadevara, who has served as a consultant to the Indian Institute of Management and to the Center for Agricultural and Rural Development at Iowa State University, is now Professor at the Center for Agriculture and Rural Development, Indian Institute of Management, Bangalore.

Ken Nicol is Associate Dean, Faculty of Management, University of Lethbridge, Alberta, Canada.

Kent D. Olson is Assistant Professor, Department of Agricultural and Applied Economics, University of Minnesota.

Orhan Saygideger, who has served as a technical advisor to USAID and World Bank projects in Somalia and to the State Planning Organization in Turkey, is now an independent consultant in project management, survey monitoring and data collection, data analysis, and agricultural sector studies.

Gary F. Vocke, formerly Adjunct Assistant Professor at Iowa State University, is now Economist at the Economic Research Service of the USDA in Washington, D.C.

James C. Wade is an Extension Specialist and Associate Professor in the Department of Agricultural Economics at the University of Arizona.

FOREWORD

IN 1958, when the late Professor Earl O. Heady founded the Center for Agricultural and Rural Development, the development and application of quantitative analytical modeling systems for support of public policy was in its infancy. The growth of CARD and of related public policy research institutes shows that the quantitative approach to analysis, design, and implementation of policy has passed the market test. In modern policy contexts it would be unthinkable to undertake a major reform without systematically and quantitatively evaluating its implications. As well, quantitative policy modeling exercises can be credited with results that have heightened public interest and stimulated public action.

The body of work summarized in this volume is an especially good example of how thorough research can lead to both advancements in science and major public policy actions. Analytical systems were developed that made it possible to evaluate the interregional and national consequences of continuing current agricultural production technologies and land use patterns. Results from these stylized but comprehensive systems pointed to a significant deterioration in the environment and the quality of the U.S. agricultural land base.

Several key natural resource and environmental policies of the 1980s were influenced by the analysis of Professor Heady and his talented students. Among these are the Food and Security Act of 1985, which provided for idling 40 million acres of land in a conservation reserve, sod-buster, swamp-buster, and other more environmentally targeted provisions, and the initiation of conservation compliance regulations linking commodity program subsidies with erosion control.

The 1990 Farm Bill will endorse the idling of frail lands and the targeting of conservation measures to preserve water quality. These are examples of major policy impacts of the research program at CARD in the area described in this book. These federal initiatives have been followed by state agricultural and environmental legislation. In 1989, for example, a total of 101 bills were introduced in 23 states on agriculture and groundwater and surface water quality.

The methods and approaches used in this research program have had broad implications both for public policy and for the importance of quantitative modeling methods. A pivotal feature of these systems was the integration of biological and physiological information in policy analysis models. This work has carried forward.

Current CARD modeling systems directly involve plant growth, soil erosion, chemical fate and transport, along with other models describing processes conditioning environmental outcomes. The economic modeling systems have been extended to an international context. More modern theoretical and computational methods are employed. But the approach represented in the land conservation and environmental improvement modeling systems summarized in this volume has had a major impact on the CARD program and public policy research.

It is a credit to Gary F. Vocke and his contemporaries that they have documented the modeling systems, the approaches, and the results from this CARD project. Perhaps their modesty results in their not emphasizing the important role that this work has played in modern environmental policy.

The scientific and policy impacts of this work are a tribute to Professor Earl O. Heady and his students and to the vision of the agencies, foundations, and other supporters of this basic and fundamental policy research program. Increasingly, public policy decisions will be based on results of specially designed quantitative analytical systems. The history documented in this book provides an example of the importance of continual investments in forward-looking modeling systems and the incorporation of advanced techniques and methods in public policy analysis.

STANLEY R. JOHNSON
Director, Center for Agricultural and Rural Development
Iowa State University

PREFACE

THE NATIONAL CONCERN over agricultural resource conservation has been intense for several decades. A peak of concern was reflected through the warnings of H. H. Bennett and the creation of the Soil Conservation Service in the 1930s. While the concern seemed to slacken during the surplus periods of the 1950s and 1960s, it returned during the 1970s. One basic cause of concern was the set of technologies adopted over several decades, which impacted on numerous facets of agriculture. The use of larger tractors and machinery units caused conservation practices such as terraces, contouring, and strip cropping to sometimes be a hindrance to field operations. The development of chemical fertilizers, which substituted for legumes in the rotation as a source of nitrogen, and other chemicals, which replaced rotational crops in control of insects and weeds, also encouraged runoff and soil erosion. In many parts of the nation farmers turned to continuous row cropping systems, which approach monoculture and result in increased soil erosion. The shift from supply control programs to a free market and the high real prices for many farm commodities also encouraged use of more marginal land and a more intensive agriculture during the 1970s.

Many environmentalists raised concerns over the agricultural technologies which began to limit wildlife habitats and pollute streams and other water bodies. Agriculturalists and conservationists also began expressing their concern over an apparent increase in soil erosion, which stood to lessen land productivity in the long run, and the increasing rate with which underground water supplies were being used. Of course, confounding conditions prevail between technology and resource depletion. Even where agricultural land's inherent productivity might have been decreased already by soil erosion, new technologies could mask this yield effect. Questions of soil erosion, soil formation, and technology are extremely complex. Some technologies intensify erosion; others provide conditions for its reduction. Some market measures place a premium on soil and water conservation; others favor their more rapid depletion. There is considerable uncertainty about the rate of soil formation and the extent to which we are turning to production methods that are not sustainable in the long run. These fears and uncertainties have caused the public to create institutions and undertake investigations capable of telling us the possibilities and consequences of resource depletion and environmental impairment. Examples from the last two decades are the

Environmental Protection Agency (EPA) and the Resources Conservation Act (RCA). The RCA passed by Congress in 1977 specifies that the U.S. Department of Agriculture and the public should make periodic assessments of resource conservation and productivity prospects in relation to energy, food supply, environmental technology, and other conditions.

To an extent, these evaluations can be made only by extremely detailed models using data that reach out to all of the land groups, water bodies, and regions of the nation. This book includes models and analytical methods which have been developed over time for these purposes. The models and their development are explained and the results of different policies and alternatives analyzed with them are discussed.

The book presents both mathematical programming and econometric models and their results as applied to specific resource problems. To keep the volume short, only a few econometric versions are reviewed. In our own approach to research, we do not concentrate on one approach or favor it over another. Our emphasis is on the use of types of models that are most appropriate for particular purposes. For most of the problems analyzed in the book, the problem posed is: "What might be the potentials, possibilities, restraints, and conditions if we try programs in the future which have not been tried in the past?" In many such cases, observations from the past are lacking and time series data do not exist for predicting the future.

A very large investment in manpower, computer funds, and general expenses has gone into the analyses and models reported in this volume. In publishing it, we are retaining the knowledge of these efforts for future science and scientists.

THE AUTHORS

ECONOMIC MODELS OF AGRICULTURAL LAND CONSERVATION AND ENVIRONMENTAL IMPROVEMENT

INTRODUCTION

by Gary F. Vocke and Earl O. Heady

UNITED STATES FARMERS have a remarkable record over many decades of steadily increasing the supply of food and fiber for U.S. consumers. The increases in output have greatly exceeded domestic requirements, allowing large exports of agricultural products to other countries. This production expansion has resulted especially from technological changes, which have transformed American agriculture from a collection of mostly self-sufficient farms in the 19th century to a highly efficient, highly mechanized, and highly specialized production system. However, many believe that this expansion of agricultural output has been attained, in part, through depletion of agricultural land and water resources. One question asked repeatedly is: Will there be adequate land and water resources to allow U.S. agriculture to continue to meet food and fiber needs at a reasonable cost to future consumers? The adequacy of land and water resources to contribute to future economic development is a continuing concern. Additionally, agriculture's impact on the environment has come under close scrutiny. Agriculture interacts significantly in many ways with the environment. Practices such as crop rotations, tillage operations, fertilizer and pesticide applications, and confined livestock operations affect the environment via runoff, sediment, nutrients, and toxic chemicals. Public policymakers and their advisors have been asked to consider legislating various kinds of regulatory schemes to conserve land and water resources and to alter agriculture's impact on the environment. To make informed decisions about proposals to control soil erosion, to limit the use of pesticides, etc., policymakers require the perspective that can only be provided by appropriate quantitative policy analyses. This book provides a review of such quantitative analyses carried out over more than a decade by Earl O. Heady and his associates with the Center for Agricultural and Rural Development (CARD) at Iowa State University.

The initial project, under which the analyses reported here were developed, was supported by the National Science Foundation's program Research Applied to National Needs (RANN). Of the several models that were financed in the early 1970s under the RANN programs, the CARD models are the only set that has continued in use. The models have been applied to numerous problems at the request of various public bodies and institutions. These models were used for the main quantitative background study for the National Water Commission. The models also were used at the request of the Environmental Protection Agency to evaluate the national environmental impacts of soil erosion and, at the request of numerous federal agencies, to make the National Water Assessment analysis for the Water Resources Council in 1974. The models were applied to an analysis of prime land use in the North Central Region at the request of the Midwest Governors Council. At the request of the Soil Conservation Service and the Economic Research Service of the U.S. Department of Agriculture, the models were used in the 1980 evaluation for the Resources Conservation Act (RCA). The models also have been used for many specific analyses at Iowa State University, other universities, and institutions and agencies such as the Argonne National Laboratories and the U.S. Forest Service. They have been used on a collaborative basis with the International Institute of Applied Systems Analyses (IIASA) in Austria and as a guide for developing parallel models in the Association of Southeast Asian Nations (ASEAN). Earl O. Heady and his associates at CARD entered into a five-year project with the Soil Conservation Service to update and respecify the models for the 1985 RCA evaluation.

This book reviews several of these analyses. It also traces the evolution of the models from those set in a mathematical programming framework only, to those which incorporate endogenous demand quantity determination, econometric components, and recursive hybrid simulation features. Rather than strike a "religious pose," which admits use only of econometric or operations research models, our approach has been to use the models that are most appropriate for the purpose at hand. Due to lack of space, we have not been able to detail all of the national and regional econometric models that have been put to use or combined with programming models. Some of these are detailed in Heady (1983).

Each of the following chapters summarizes a study from this continuing analytical effort to model agricultural production and its interaction with resource conservation and the environment. An overview of the studies included in this book is provided at the end of this introduction. The remainder of the introduction provides a brief review of features of the U.S. agricultural situation that are especially relevant as background for these studies, summarizes evolution of models for the purpose at hand, and provides an introductory discussion of the conceptual framework and analytical procedures employed in these analyses.

AGRICULTURAL PRODUCTIVITY

The total land used for crops has fluctuated from a high of 380 million acres just after World War II to less than 333 million acres in the late 1960s and again in 1983 when large acreages were idled by government programs (Table I.1). Although total land used in 1979 was the same as in 1929, considerably more output was produced in 1979. The data in Table I.2 illustrate both the changes in the use of land and other agricultural inputs and the changes in productivity. The total quantity of inputs employed in agriculture has increased only slightly since 1920. However, the improved quality of agricultural inputs and substitution among inputs resulted in total factor productivity (changes in output obtained

Table I.1. Major uses of land for selected years, 1924-81

Year	Cropland Harvested	Crop Failure	Fallow	Total Used for Crops	Idle	Total Cropland Excluding Pasture	Acres Idled By Govt Programs
	(million acres)						
1924	346	13	6	365	26	391	0
1929	356	13	10	379	34	413	0
1934	296	64	15	375	40	415	0
1939	321	21	21	363	36	399	0
1944	353	10	16	379	24	403	0
1949	352	9	26	387	22	409	0
1954	339	13	38	380	19	399	0
1959	317	10	31	359	33	392	22
1964	292	6	37	335	52	287	55
1969	286	6	41	333	51	384	58
1972	289	7	38	334	51	385	62
1973	316	5	31	352	32	384	19
1974	322	8	31	361	21	382	3
1975	330	6	30	366	NA[a]	NA	2
1976	331	9	30	370	NA	NA	2
1977	338	9	30	377	NA	NA	0
1978	331	7	31	369	NA	NA	18
1979	342	7	30	379	NA	NA	12
1980	341	12	31	384	NA	NA	0
1981	353	17	31	391	NA	NA	0

Source: United States Department of Agriculture, 1983.

[a]NA: Not available.

from all inputs) increasing 135 percent since 1920. The productivity of land, measured as crop production per acre, more than doubled.

In recent years, a relatively large proportion of the increased output of agricultural crops represented by the index in Table I.2 has been exported. These exports are an important issue in the ongoing discussions about resource adequacy and environmental concerns. In addition, exports of agricultural commodities are an important variable influencing commodity prices and farm incomes and the foreign exchange earnings of the U.S. economy.

World trade in grains increased in the 1970s in response to rapid economic growth, an expanding world population, higher per capita incomes, and increased consumption of livestock and poultry products. During the decade, combined world exports of wheat, rice, and coarse grains rose at an annual average rate of 5.1 million metric tons (Table I.3).

The U.S. share of this expanding world trade in grains increased markedly during the 1970s. Exports of wheat, cotton, soybeans, and corn are very important for U.S. agriculture. More than one-half of the total U.S. production of wheat, soybeans, and cotton recently has gone into foreign markets (Table I.4). More than 90 percent of these exports are commercial sales. As a result of these greatly expanded sales of U.S. farm products overseas, the net surplus in U.S. agricultural trade expanded sharply during the decade, partially offsetting a sharp deterioration in the U.S. nonagricultural trade balance. Thus, exports of farm products played an important role in helping to pay for the nation's large imports of petroleum and other materials (Table I.5).

The cropland base on which these supplies of agricultural commodities have been produced has been inventoried periodically. In the 1977 National Resource Inventories (NRI) conducted by the U.S. Soil Conservation Service (SCS) as part of the Resource Conservation Act (RCA), the 1.5 billion acres of nonfederal land in the United States were classified as follows (Council for Agricultural Science and Technology, July 1981):

413 million acres of cropland (27 percent);

Table I.2. Index measures (1977 = 100) of resource use, output, and farm productivity, 1920-81

	Selected Inputs				Output			Productivity (Output/Input)		
Year	All inputs	Labor	Real estate	Mechanical power and machinery	Livestock	Crops	Total	All inputs	Land[a]	Labor
1920	96	486	105	27	42	50	42	44	52	8
1930	99	465	104	34	51	45	43	44	44	9
1940	97	417	106	36	57	51	50	52	53	12
1950	102	310	108	72	70	59	61	60	59	19
1960	98	207	102	83	82	72	76	77	77	37
1970	97	126	104	86	99	77	84	87	88	66
1971	98	123	102	87	100	86	92	94	96	74
1972	98	117	101	87	101	87	91	94	99	78
1973	98	115	100	90	99	92	93	95	99	81
1974	98	112	98	93	100	84	88	90	88	79
1975	97	107	97	96	95	93	95	99	96	89
1976	99	103	98	98	99	92	97	97	94	94
1977	100	100	100	100	100	100	100	100	100	100
1978	101	95	100	104	101	102	104	102	105	109
1979	104	93	100	107	104	113	111	106	113	119
1980	103	92	101	105	108	101	103	100	99	112
1981	103	90	101	105	108	117	116	112	114	129

Source: United States Department of Agriculture, 1983.

[a]Measured as crop production per acre.

Table I.3. Annual change in total world grain trade, 1971-72 through 1983-84

Marketing Year	% Change from Previous Year in Total Trade
1971-72	+0.2
1972-73	+24.7
1973-74	+7.1
1974-75	-5.0
1975-76	+13.5
1976-77	+6.5
1977-78	+9.7
1978-79	+7.8
1979-80	+25.0
1980-81	+12.4
1981-82	+4.8
1982-83	-8.7
1983-84	+3.0

Source: Wisner and Denbaly, 1983.

Table I.4. Percentages of major U.S. crops used domestically or exported for the market years of 1969-70 and 1980-81

	1969-70		1980-81	
	Exports	Domestic use	Exports	Domestic use
Wheat	44	56	66	34
Cotton	26	74	50	50
Soybeans	49	51	55	45
Corn	13	87	33	67

Source: Wisner and Denbaly, 1982.

Table I.5. Growth in U.S. agricultural trade

	1950		1972		1981		Compound Annual Growth Rates 1950-81	
Item	Volume	Value	Volume	Value	Volume	Value	Volume	Value
	1,000 metric tons	Million dollars	1,000 metric tons	million dollars	1,000 metric tons	million dollars	Percent	
Grain and Products	11,764	1,034	45,362	2,875	117,200	19,708	7.7	10.0
Oilseeds and Products	1,019	168	15,931	2,137	44,151	10,427	12.9	14.2

Source: United States Department of Agriculture, 1982.

548 million acres of pasture and rangeland (36 percent);
377 million acres of forest land (25 percent);
175 million acres in other uses, including urban and developed areas, highways, and airports (12 percent).

In addition to the 413 million acres of land classed as available for crop production, 127 million more acres of pasture, range, and forest land are considered to have high potential for conversion to cropland use with little investment. Another 91 million acres have medium potential but would require some development costs and conservation investments to remain in crop use. A regional breakdown of total dryland and irrigated cropland, as well as land with potential for conversion to cropland, is presented in Table I.6.

In some areas of the United States, a critically important issue is the depletion of the present agricultural land base through erosion. As the total acreage of crop production expands in the future, potential erosion problems can be expected to become more severe, including environmental problems arising from the associated sedimentation of the nation's waterways. Most erosion losses are the result of water runoff, estimated in 1977 to be 4.044 billion tons, the equivalent of 2,247,000 acre-feet of soil (United States Department of Agriculture, 198 lb). (An acre-foot is a one-foot-deep slice of soil large enough to cover one acre.) A second source of erosion is wind, responsible in 1977 for the movement of 1.462 billion tons (812,000 acre-feet).

Table I.6. U.S. cropland and potential cropland by region

Regions[a]	Dryland and Irrigated Cropland	Potential for Conversion to Cropland	
		High potential[b]	Medium potential[c]
	(1,000 acres)		
Northeast	16,916	1,079	3,683
Appalacian	20,753	4,737	9,806
Southeast	17,506	4,924	10,922
Lake States	44,141	2,287	6,295
Corn Belt	89,922	4,841	9,670
Delta States	21,191	3,092	6,989
Northern Plains	94,574	5,050	12,845
Southern Plains	42,222	5,217	14,846
Mountain	40,912	3,234	11,056
Pacific	23,172	1,631	3,920

Sources: United States Department of Agriculture, 1981.
United States Department of Agriculture, 1980.

[a]Northeast includes Connecticut, Delaware, Maine, Maryland, Massachusetts, New Hampshire, New Jersey, New York, Pennsylvania, Rhode Island, and Vermont. Appalachian includes Kentucky, North Carolina, Tennessee, Virginia, and West Virginia. Southeast includes Alabama, Florida, Georgia, and South Carolina. Lake States include Michigan, Minnesota and Wisconsin. Corn Belt includes Illinois, Indiana, Iowa, Missouri, and Ohio. Delta States includes Arkansas, Louisiana, and Mississippi. Northern Plains includes Kansas, Nebraska, North Dakota, and South Dakota. Southern Plains includes Oklahoma and Texas. Mountain includes Colorado, Idaho, Montana, Nevada, New Mexico, Utah, and Wyoming. Pacific includes California, Oregon, and Washington.

[b]These lands are considered to be convertible to cropland use with little investment.

[c]These land are considered to be convertible to cropland use with some development costs and conservation investments to remain in crop use.

The following areas are examples within major regions of the United States where erosion is especially severe, along with some of the reasons for the excessive erosion (Council for Agricultural Science and Technology, 1982).

North Central States

Soil erosion is especially serious in the North Central region of the United States. The long-term consequences of this situation are important because the twelve states in this region produce 45 percent of the food grains, 77 percent of the feed grains, 64 percent of the oilseeds, and 45 percent of all livestock sales in the United States. The heart of this region is the Corn Belt, one of the richest agricultural acreas in the world. NRI data indicate that about 43 percent of the rowcropped land in the Cornbelt is highly subject to erosion. Erosion losses of 5 tons or more per acre annually are common in this region, and losses of 10 tons or more per acre occur on 19 percent of the rowcropped land.

Southeastern States

Rainfall in the Southeast is high and runoff and excessive erosion often occur. According to the NRI survey, erosion rates of more than 11 tons per acre per year occur on about one-third of the land used for row crops in the twelve southwestern states. The silty uplands of the southern Mississippi Valley, which cover about 26,000 square miles in western Kentucky, Tennessee, Mississippi, and parts of Louisiana and Arkansas, have cropland erosion rates that are among the highest in the United States. The deep, wind-deposited materials from which the soils of the area are developed are very productive but easily eroded. Row crops are grown extensively, largely without conservation practices. Soil losses from erosion average up to 20 tons per acre per year. The average annual soil loss in this region is 9.5 tons per acre per year. Excessive erosion has made crop production uneconomical in some areas of this region.

Mountain States

In the Central High Plains and the plateau lands east of the Rocky Mountains, winter wheat and sorghum are grown extensively and corn is grown with irrigation. Both wind and water erosion are extensive on these croplands. In Colorado, New Mexico, and Texas, the average annual wind-erosion rates on cropland are 9, 11, and 15 tons per acre per year, respectively. When compared with the total acreage suffering excessive erosion in the Corn Belt, the excessive erosion acreages for Colorado, New Mexico, and Texas are small.

Pacific States

The Palouse and Nez Perce prairies and parts of the Columbia Plateau in Washington and Oregon experience massive sheet and rill erosion. Slopes in this area are often very long and steep, and a high proportion of the land is used for crops. Most of the erosion is associated with runoff from low-intensity rainfall and snow melt on partially frozen ground. Parts of these prairies and plateau regions with significant slopes and relatively high rainfall have average erosion rates of 15 to 25 tons per acre per year.

Because of the many variables involved, the effects of soil erosion on soil productivity cannot be assessed with precision. Historically, these effects have been masked by the increased use of technological inputs in crop production and steadily improving crop varieties. There is considerable research, however, which suggests that excessive erosion significantly impairs productivity and that erosion has increased input costs (Alt et al., 1981).

While the impact of soil erosion on the future productivity of the agricultural land resources is a major problem, it is not the only important problem associated with erosion. Eroded soil may clog the nation's waterways, fill in flood control structures, impair wildlife habitats, etc. The public concern for improvement of water quality as affected by agriculture has focused on controlling erosion and runoff from agricultural areas and on the use of pesticides, which may be carried into the waterways with the eroded soil. Concern about water quality has also resulted in major efforts to control runoff from livestock feedlots that may carry livestock wastes into waterways.

The quantity as well as quality of the water resources of the United States is important to the production capacity of agriculture. Agriculture is the largest user of water in the United States. Almost one-half of all withdrawals from surface and ground sources are for U.S. agriculture. Almost 98 percent of the water used in agriculture is for irrigating crops. The remainder is for livestock.

The production capacity of U.S. agriculture has been significantly enhanced by irrigation since the early 1900s. In 1900 there were 7.17 million irrigated acres; by 1980 54.1 million acres were being irrigated. The use of the nation's water supplies has increased sufficiently in some areas to cause a concern about groundwater depletion and surface water shortages. One of the most critical groundwater depletion areas in the United States is the High Plains of eastern Texas and eastern New Mexico. Competition for existing surface water supplies is intense. Energy development and municipal and industrial uses are bidding away water previously used for agriculture or recreation or to maintain fish and wildlife habitats. In the absence of alternative water supplies or improved irrigation efficiency, irrigated land may have to be converted to dryland production. At the national level, the production losses due to the abandoning of irrigation will have to be made up elsewhere if output is to be maintained. This could aggravate the soil erosion problem if additional lands subject to erosion in humid regions are brought into production.

GOALS AND OBJECTIVES FOR THE STUDIES

Research relating to the analysis and planning of land and water use and environmental quality typically fall into one or more of the four following major categories:

1) Inventory and descriptive studies, which quantify the amount and characteristics of land and water resources and usually suggest restraints on resource use in terms of upper limits on availability or physical suitability to various uses.

2) Predictive (usually statistical) studies, which explain behavior in the use of land and water as it relates to economic, social, policy, and other variables.

3) Normative allocation models, which determine how land and water ought to be used relative to variables and parameters that relate potential use to economic and institutional restraints while considering various goals or objectives. These models also can be used to indicate potential production, soil erosion, water use, resource values, and related items under programs that have not been tried but that might be initiated in the future.

4) Simulation models, which predict outcomes when exogenous variables affecting land and water use are assigned certain values and endogenous variables are determined within the estimated relationships of the model(s).

Both the programming and the simulation models can be used to indicate outcomes when different policies or futures are to be examined or posed. These modeling efforts usually involve activities from all four categories. For example, descriptive studies of land resources are used to determine the land restraints in a programming model. The predictive studies can provide demand functions for the objective function of a nonlinear programming model or provide behavioral equations for a statistically based simulation model. Programming models are used when there are no time series from the past upon which econometric or statistical models can be based. Also, programming models can be developed and applied with inventory data for hundreds of producing regions and thousands of small land classes for which time series data for econometric analysis is impossible. Planning often must be done on the basis of data and intelligence accumulated for these purposes, even though the planned event has never happened in the past. For example, we may wish to determine potential crop production if soil erosion were to be reduced. Or we may wish to estimate the impact on U.S. agricultural productivity if certain pesticides were banned. Programming models can usually provide such estimates for hundreds of producing areas and thousands of land groups, while statistical models cannot. Econometric models, however, can be combined with programming models to estimate the price effect of these potential outcomes. These possibilities have been shown in Langley, Huang, and Heady (1981) and Schatzer and Heady (1982).

Most of the models summarized in this book are programming models and indicate optimal spatial allocation of agricultural production relative to restraints in markets, natural resource supplies, environmental quality controls, alternatives in domestic consumption, export policies, etc. These models and the results obtained are developed under research projects at CARD within the conceptual framework of interregional competition analysis. However, while they are not explained in detail, a series of econometric models aid in these purposes (Heady, 1983).

CONCEPTUAL FRAMEWORK FOR INTERREGIONAL COMPETITION ANALYSIS

The economic principles that provide the conceptual framework for the studies of interregional competition in this book are the principles of specialization and comparative advantage. Under the principle of specialization, different regions have unique combinations of natural resources and are suited in varying degrees to the production requirements of agricultural commodities. In addition, by investments in various kinds of improvements these resources can be further adapted to the production needs of certain commodities. Most agricultural areas, however, produce more than one agricultural commodity because of complementary and supplementary relationships. Examples of complementary relationships are the contributions of the production of feed-grain crops to livestock production and the plant nutrients in livestock manures to crop production. Examples of supplementary relationships are the better utilization of resources such as labor when different commodities have different seasonal requirements.

The principle of specialization of itself, however, does not explain what agricultural commodities will be grown in each region. The principle of comparative advantage is needed to complete the conceptual framework needed for the study of interregional competition. A region may be capable of producing an excellent soybean crop and yet be actually used for growing only a reasonably good crop of wheat. Such a result may be obtained because there is a relative abundance of land well suited for growing soybeans and a relatively small amount of good wheat land (in both cases, of course, relative to the world's needs for these crops). Thus some land that might grow soybeans well is planted to wheat, although it is only fairly well suited to wheat. In this way the world gets enough wheat and not too many soybeans. If the world's demand for soybeans were to increase while the demand for wheat remained the same, some of this land would be diverted away from wheat to soybeans. For example, if the world's demand for beef were to increase significantly, land would be diverted to soybeans so that needed soybean meal would be available.

ANALYTICAL APPROACHES

To make projections of future demands on and requirements for U.S. agricultural resources requires analytical tools that generate empirical results at both national and regional levels. National detail is required so that market impacts on prices can be measured, supplies can be equated with demand, and interdependence among regions can be established. Detail by region is needed so that restraints on production and resource use can be measured and evaluated. This detail is especially important for analyses related to equity in the impact of alterations in technologies and resource availabilities prescribed by national policies.

Four analytical approaches to estimating the impacts of policy changes will be discussed briefly to provide a perspective on the modeling efforts reviewed in the following chapters. These analytical approaches are partial budgeting, input-output modeling, econometric simulation, and mathematical programming. All four approaches are capable of facilitating empirical analysis. The quantitative results obtained, however, are reliable only to the extent that accurate judgments and assumptions are entered and the analytical approach is appropriate to the problem being investigated. The advantages and disadvantage of these four selected approaches for the study of land and water use within the framework of interregional competition for policy analysis are discussed briefly in the following paragraphs.

Partial Budgeting

The partial budgeting approach uses the least sophisticated mathematics and perhaps is the most easily understood. With partial budgeting one must specify region by region how much acreages, yields, production costs, and other relevant factors will change as the result of production and consumption responses to policy changes. The procedure begins by developing budgets for representative farms in each region reflecting the organization of farms under current conditions and then how these farms might be reorganized under the proposed policy changes. These representative farm budgets are then aggregated to the regional level to obtain the measure of the interregional consequences. However, the partial budgeting approach has severe limitations for the analysis of interregional consequences of policy. The complexity of the effects of changes in comparative advantages among regions is difficult to comprehend and substitutions between inputs and intermediate products within a region are difficult to analyze and enter in the budgets. In addition, the aggregation of the completed budgets to the regional level in an unbiased manner is not an easy task.

Input-Output Modeling

The input-output model provides an operational framework for concisely quantifying activity among economic sectors. The input-output model allows a comprehensive picture of the interrelationships between regions and sectors. The direct effects, as well as the indirect effects, of final demand changes can be investigated. The model is useful for descriptive purposes in viewing certain interrelationships among spatial sectors of agriculture.

Although the input-output model is a convenient tool for making projections, these projections have limitations because of the assumptions underlying the model. These assumptions include constant input-output coefficients, zero rates of substitution, and fixed relationships both within and among sectors. In this context, the input-output model is very useful for describing the interdependence among agricultural regions, but not for expressing the effects of economic forces created by improved technology or the effects of lower factor prices in one region on output, prices, and income in another region. Generally, the input-output model expresses the complementary, rather than the competitive, relationships among regions.

Econometric Simulation

Econometric simulation models for interregional competition are sets of equations that reflect the structure of supply and demand. The coefficients of these equations quantify past behavior and associations among variables. The regional supply and demand relationships are specified using economic theory, and the functions are then estimated using historical data. The market behavior assumption implied by this procedure is that the average of past responses will allow the prediction of future response.

Econometric simulation models generate period by period estimates of the values of the endogenous variables. (Endogenous variables are variables whose levels are determined in the model.) There are many different types of econometric models. For illustrative purposes, a recursive model will be used as an example of econometric simulation. To employ a recursive econometric simulation model, one uses (1) initial conditions (usually data from a recent year), (2) the assumed future values of variables which are exogenous to the model (those variables whose values are determined outside the model), and (3) the model relationships to simulate a time path. Interregional competition analysis using econometric simulation typically involves shifting regional supply curves appropriately to reflect the impacts of policy changes. For example, supply equations may be shifted and the slope or responsiveness may be adjusted. The resulting change in the simulated time path of the variables from the original result is the estimate of the economic effect of the policy change.

Mathematical Programming

Mathematical programming models designed to develop the economic forces of interregional competition implied in the production possibilities of agriculture possess the common structure of an objective function to be maximized or minimized subject to a system of restraints. The objective function is maximized or minimized by determining the level for the decision variables or activities of the model. These activities are the processes by which agricultural resources are transformed into crop and livestock products and then transported to the consumer. Assumptions and judgments are involved in specifying the restraints on availability of resources, the rapidity with which interregional production shifts can occur, etc. The programming mode calculates the best agricultural program to optimize its objective function. This mathematically optimal result shows what agriculture ought to do to optimize the objective function, which might be, for example, to minimize the cost of producing food and fiber to meet the nation's food needs, to attain a given level of soil erosion reduction at least cost, etc. Programming models have an advantage over input-output models and econometric simulation in developing the impacts of policy changes that are aimed at greatly changing (1) the mix of products produced, (2) the regional proportion of, commodities produced (3) the relative mix of resources used, and (4) other very detailed structural relationships.

Modeling of Interregional Competition Using Mathematical Programming

The modeling of interregional competition in U.S. agriculture using mathematical programming begins with the delineation of the regions for the model. The interregional models in the studies in this book are based on three types of regional delineations: producing areas, water supply regions, and market regions. The producing areas are disaggregated into land quality groups based on variations in land quality.

The following discussion develops the role in the programing model of each of these regions with respect to the availability of land and water resources, crop and livestock production, the interregional transportation network, and the domestic and foreign demands for the agricultural commodities. This is followed by a description of the mathematical properties of the solution from a programming model and the solution's correspondence to the long-run equilibrium of the agricultural sector for policy analysis.

Delineation of the Regions for Crop Production

The agricultural land available for cropping and grazing in the model is defined within regions referred to as producing areas. These producing areas each encompass an agricultural area of relatively homogeneous land resources and farming practices. Because of the variability of agricultural lands within the producing areas in terms of productivity and susceptibility to erosion, the land resources are categorized into one of several land groups. Cropping systems, which account for the productivity of the land, the cost of the other factors of production such as pesticides, machinery, etc., and the relationship between management and tillage practices and soil erosion, are defined for each land group within each producing area consistent with the year in the future for which the projection is being made (for example, the year 2000). These cropping systems (referred to as activities in the models) include the production of crops that may not be the best suited for that region, but are feasible in that region. This procedure of defining all feasible cropping systems allows for a better analysis of interregional competition, as discussed earlier in this chapter. In some of the interregional competition studies for the United States we have divided the country into 223 producing areas. In other studies, we have used 105, 150, 90, or any other number most appropriate for the study. Also, the number of land groups included per producing region has varied from 5 to 12.

In the western part of the nation, water, as well as land, is a potentially scarce resource. Thus water supplies available for irrigation and livestock are defined within regions, referred to as water regions, in the models for the western United States. These water regions are aggregations of contiguous producing areas. For those producing areas contained within these water regions, the land resources are classified as to availability for irrigated crop production, and cropping systems (or activities) for irrigated crop production are defined for these lands.

Resource Base for the Programming Models

The purpose of this section is to describe two alternative ways of defining the land resource base of the programming models. One approach is the aggregation of representative farms within a region. A second approach is the regional aggregation of agricultural resources into producing areas. The resource base of the models developed in this book are formulated using a regional aggregation of resources. By contrasting the regional-aggregation-of-resources-into producing-areas approach with the aggregation-of-representative-farms approach it is hoped that a better understanding of the models can be obtained. The advantages and disadvantages of these two approaches are discussed briefly in the following paragraphs.

Aggregation of representative farms. This procedure involves classifying all the farms in a region into one of several farm-type groups and designating a representative (typical) farm for each group. Production restraints and activities are specified in a programming model for each representative farm and an optimal production plan is determined. An optimal plan for the whole farm-type group is then obtained by multiplying each optimal representative farm plan by the number of farms in its respective group (or some other appropriate weighting factor). The regional total is then the sum of the group totals. The process is repeated for each region. This type of model has been used to analyze regional supply response potentials. The approach is difficult to employ because of the complexities of aggregating the responses of the representative units. Certain aggregation biases arise unless specific conditions are met. To analyze regional resource allocation in the sense of jointly determining production levels in many interdependent regions, each representative farm has to be treated as a region with its own restraints and production activities within a single interregional programming model. Thus some resources could be unlimited for individual representative farms but fixed for the aggregate of all such farms. The final solutions for each representative farm or resource situation subaggregate would reflect the competition for the fixed amount of resources available at the regional level. The validity of results so obtained depend on the extent to which resources are mobile among regions. For example, the supply of land may be fixed for a region, but labor and capital may flow among regions. The data and computer requirements of such an approach are considerable.

Regional aggregation of resources. An alternative approach to formulating the resource base for an interregional programming model is to directly use the producing area as the unit for defining the relevant resource restraints. Agricultural resource restraints (typically land and water) and crop-producing activities are then defined for each producing area. An important feature of this approach of aggregating resources within producing areas is that of the implied normative long-run adjustment possibilities of production patterns for regions. Some variables that are extremely important in the short run are omitted and thus assumed perfectly flexible in the long run. For example, regional restraints on available labor and capital are not included (although they could be) in these models, as these variables are assumed to be completely mobile between the present and the year of the projection. These and similar assumptions are made to allow investigation of more fundamental long-term relationships.

Another important feature associated with the aggregation of resources to the producing areas is that the variability in natural endowments in soil, water, and climate is aggregated into an "average" set of technical coefficients. Consequently, whole regions might make complete shifts from the production of one crop to another crop in the model in a way that does not reflect the reality of American agriculture. Specialization resulting from regional comparative advantage occurs within the agricultural sector, as evidenced by the concentration of corn in the Midwest, cotton in a more limited area of the South and West, soybeans in the Midwest (and more recently in the South Central states), and wheat in the Great Plains. Specialization by regions is a function of spatial price patterns as reflected in transportation costs and demand concentration, relative yields as affected by soils and climate, local resource supplies and prices, and the relative profitability of different crop and livestock enterprises for the individual farmer. However, regions do not specialize completely because of the production advantages resulting from mixed farming patterns. Among the advantages of the multi-activity farming pattern are timeliness, risk aversion, cropping patterns consistent with resource management, complementarity expressed in pest control and soil fertility, seasonal requirements for and availability of resource services, and farmers' preferences. Thus the modeler must use judgment in specifying the assumptions to be incorporated into the model concerning intraregional shifts in crop production.

Defining activities and restraints within the programming model to account for long- versus short-run adjustment possibilities and for intraregional shifts in production patterns is time-consuming and rests on the judgment of the modelers and the objectives of the analysis.

Market Regions for Agricultural Commodities

Market regions in the models are aggregations of contiguous producing areas in which the demand for agricultural products is balanced with the supply of agricultural products. The demand sectors for agricultural commodities, both domestic and foreign, are defined at the market region level. Foreign demand is defined only in those market regions containing a major grain shipment port. The domestic demand for agricultural commodities includes the livestock sector of the model, which utilizes grain, silages, hay, and pasture from the producing areas as intermediate products in the production of meat, eggs, and milk.

Agricultural production in the producing areas contributes directly to the supply of agricultural commodities in the market region in which the producing area is contained. Transportation activities permit movement of commodities to market regions with demands exceeding supply from their own producing areas. This interregional transportation of agricultural commodities allows the model to reflect in its optimal program the comparative advantage of the producing areas in the production of agricultural commodities. This important transportation link allows the substitution of resources in one producing area for resources in another producing area. For example, land in one producing area can be substituted for water in another producing area a thousand miles away (or vice versa) in the production of agricultural commodities. This detail concerning substitution of resources in production is especially important for analyses related to equity in the impact of alterations in technologies and resource availabilities prescribed by environmental or other national policies.

The Objective Function of the Programming Model

Mathematical programming models are characterized by optimizing objective functions of various formulations within bounds or restrictions reflecting the availability of resources for agricultural production and various technical, economic, social, and institutional constraints, as discussed earlier. The selection and specification of objective functions have important implications in interpreting solutions from the models. In most of the models discussed in this book a least-cost objective function is used. The least-cost objective function is meant to minimize the cost of producing and transporting the food and fiber requirements specified for both domestic and foreign consumers of U.S. agricultural commodities. The linear programming model is used for this purpose, but not other purposes, in large-scale interregional competition modeling because of both computational ease and correspondence to many aspects of the competitive functioning of the real world of U.S. agriculture when appropriately formulated. The least-cost model is also a long-run equilibrium model in the sense that it supposes all resources (except land and water) receive market rates of return. Returns to land and water are determined endogenously.

Utilizing the cost and output data incorporated in the crop and livestock activities and the cost of transporting agricultural commodities between market regions, the solution of this model specifies in each market region the mix of crops to be raised in each producing area, which cropping systems will be employed, the conservation practices to be used, and, in some of the models, the livestock to be raised and their rations. The solution also includes the interregional transportation of agricultural products. All of these results are obtained in accordance with the principles of specialization and comparative advantage, as developed earlier. In addition, the solution generates shadow prices for the commodities and for the land and water resources which are useful in analysis of the results.

However, in an interregional linear programming model incorporating a least-cost objective function, it is important to recognize that a demand analysis is carried out beforehand, and the quantities of each of the agricultural products projected to be required for the year being analyzed (for example, the year 2000) are incorporated into the model as specific restraints in each of the market regions. The consequence of this procedure is that the model does not take into account that consumption levels of individual agricultural commodities depend in varying degrees on the price of that commodity and other commodities. These aspects of demand analysis are explained in more detail in the chapters in this book concerned with quadratic and separable programming models, which are formulated to account for this interaction between consumption levels and price.

Use of Programming Models for Policy Analysis

Once it has been determined that a national interregional competition programming model is the appropriate analytical tool for the issues to be investigated, the model must be properly formulated. The model must account for (1) the land and water resources available to agriculture, (2) crop and livestock production activities for the transformation of these resources into agricultural commodities, (3) the commodity transportation network, and (4) the domestic and foreign demands for agricultural products. The appraisal of soil, water, and related agricultural resources is also a part of the activities that must be carried out in the development of the programming models employed in the analyses in this book. This appraisal must allow for the development of information about actual and potential regional relationships in the agricultural sector. The data and relationships developed from the appraisal are used in structuring and specifying appropriate interregional competition programming models.

The resulting programming models can then be used to make projections of the consequences of changes in national and regional policies affecting agriculture. By making appropriate modifications, the models can be used for evaluating policy changes affecting (1) regional resource availabilities, (2) soil loss or other environmental limits, (3) fertilizer usage, (4) commodity demand, (5) the export market for commodities, (6) farming techniques, and (7) supply controls or marketing quotas.

Regional resource availabilities could be altered by policies affecting land use, water availabilities or transfer, and the availability of other endogenously allocated inputs such as fertilizer or pasture. Soil erosion could be affected by regulations directly limiting per acre gross soil losses or indirectly restricting certain types of farming practices. The fertilizer input levels are affected by price changes or by allowable application rates. Changes in demand, as reflected in per capita consumption levels or commodity substitutions in the diet, can be reflected by altering the regional demands. Similarly, the impact of policies affecting international trade can be analyzed by altering this portion of the demand sector, including the regional export allocation.

A broad category of policies affecting farming techniques can be evaluated by the models developed and presented. Shifts from dryland to irrigated agriculture as a new irrigation development is initiated, shifts to new tillage practices such as minimum tillage, new varieties that expand the regional compatibility of a crop, and technologies affecting the use of the commodities or the efficiency of their transformation are only a few of the farming techniques that can be evaluated using such models. Supply-control or resource-use policies can be investigated by either minimum or maximum restraints on acreage or production by regions.

The basic purpose of using models for policy analysis is to assess the effects of a changed policy situation so that predictions can be made about the likely consequences of the policy. This process involves the following steps: An optimal solution is obtained from the appropriate model, using the baseline assumptions under the existing policy situation. Next, the model is modified or respecified in an appropriate manner to reflect a change in policy or structure. For example, a changed production restraint consistent with change in export policy gives a new solution. The two solutions are then compared and any differences between solutions are attributed to the modification, i.e., the changed production restraint reflecting the policy change.

When combined with the knowledge of the modeler about the model, this comparison of the solutions provides the basis for assessing the impacts of the policy changes being evaluated. The quantitative results available on a regional basis include data on (1) acres and yields of dryland and irrigated crops by land group, (2) nitrogen utilization and price, (3) aggregate water use, (4) value of resource use in crop production, (5) the quantity and value of resource use in livestock production, (6) the quantity and value of the commodities produced, (7) land use, value of land, soil loss, and unused or idle land, (8) acreage and soil loss by conservation tillage practice, (9) acreage of land farmed under rotations of varying length, (10) water balance by water supply regions for both consumption and withdrawal needs, (11) interregional commodity flows, (12) livestock production and rations, (13) sediment production, (14) commodity supply price, and (15) a range of other interesting economic and physical data.

The results of the programming and related models are not necessarily predictions of what will happen in the future. Rather, the results are projections useful for assessing the possible or potential impacts of variations in policy. These projections are dependent on the assumptions from which they are derived and thus are conditional statements about the future.

OVERVIEW OF THE MODELING PROJECTS

The interregional competition studies were carried out by CARD personnel at Iowa State University. These studies were financed under various projects including the National Science Foundation's Research Applied to National Needs (RANN), the National Water Assessment (NWA) project with the Natural Resources Division, Economic Research Service of the USDA, the North Central Regional Center for Rural Development, and the Resources Conservation Act (RCA) evaluation with the Soil Conservation Service of the USDA. The latter has been especially important in the development and application of these models. A brief description of the models employed in each of the studies reviewed in the book is presented below.

CARD-RANN Model

This national interregional linear programming model formulated in 1972-74 was designed for studying relationships of agricultural production to land and water use and the environment for the year 2000. The model has 223 producing areas with the agricultural land aggregated into 9 separate land groups, 51 water regions, and 30 market regions. Commodities included endogenously in the model are barley, corn, corn silage, cotton, legume hay, nonlegume hay, oats, sorghum, sorghum silage, soybeans, sugar beets, wheat, summer-fallow, dairy cows, beef cows, beef feeding, and pork. The environmental quality aspects of the model include soil loss, nitrogen use, and animal wastes.

The model was used in a study made to evaluate impacts on the agricultural sector of imposed limits on sheet and rill erosion of cultivated lands. The main questions the study addressed were: Does agriculture have sufficient production capacity to meet domestic and export demands and also contribute to improvement of the environment through reducing erosion? If so, how far can environmental attainment through reduction of soil loss from agriculture be carried while food demands are met at reasonable real prices? What interregional changes in crop and livestock production, land use, and water allocation occur as different levels of environmental improvement through soil-loss controls are attained?

The CARD-RANN national model was respecified in 1974-75 to evaluate the various impacts when land use policies are applied in one state (in this case Iowa) but not nationally. The CARD-RANN model was modified by delineating twelve producing areas within the state of Iowa. The model developed the impacts of policies for altering gross soil erosion, nitrogen use, and pesticides in Iowa but not the rest of the country. This study was made to help determine the impact on costs and returns in agriculture in one state if it instituted a conservation or environmental improvement program while others did not.

CARD-NWA Models

This national interregional linear programming model, like the CARD-RANN model, was designed for studying relationships of agricultural production to land and water use and the environment for the nation. The basic structure of the two models was somewhat similar. The major differences between the two models was in the delineation of producing regions, market regions, and water regions, and in updated and improved technological coefficients. The NWA model, formulated in 1974-76, was used to evaluate the nation's land and water resource adequacies relative to future magnitudes and trends of variables affecting agriculture under varying assumptions of available technology and national resource policy. Particular emphasis was placed on identifying national and regional resources likely to be in critical supply in the year 2000.

This CARD-NWA model was reduced from a nine-land-group model to a five-land-group model for two additional studies reported in this book. The first study, carried out in 1977-78, used the model to evaluate the impact of national policies designed to curb pollution problems created by excessive erosion of soil, persistence of certain organo-chlorine insecticides in the environment, feedlot runoff, and nitrogen fertilizer use. Another study, in 1978-79, used a variation of the model to analyze tradeoffs between goals of production efficiency and soil-loss reduction for U.S. agriculture.

The CARD-NWA model, with major modifications in specification, also was employed in two studies concerned directly with sediment loads in the nation's waterways attributable to agricultural sources. For these studies the CARD-NWA model was modified to include a sediment-delivery sector. The addition of the sediment-delivery sector allowed the analysis of the interrelationships between sediment loads in the major river systems of the country and the additional costs for agriculture of controls on erosion and sedimentation from cropland. This CARD-NWA model, with the sedimentation sector specified for it, was used in a study investigating national policies concerning the control of the sedimentation of river systems from agriculture and in a study comparing the various impacts of limiting agricultural sediment through regulations imposing limits on gross soil erosion with a taxing program on sediment delivered to the main river basins.

CARD-RCA Model

This national interregional linear programming model, formulated for the 1980 RCA evaluation and continuing for the 1985 evaluation, is similar to the CARD-NWA model except for two important features: the livestock sector was exogenous to the model, and the crop sector was modified to allow the adjustment of crop yields for soil erosion losses. The study employing this respecified model included an analysis of the impact of allowing the development of potential cropland (as defined earlier in the NRI) in conjunction with a range of projections for technological improvement for the year 2030. Also included was an examination of the impacts on U.S. agriculture of several levels of soil losses for the year 2030 under a moderate supply-moderate demand scenario.

Nonlinear Programming Models

Fixed-demand national linear programming models do not fully account for the interaction of supply/demand relationships in the determination of national prices and production. Two nonlinear programming models which incorporate demand relationships, rather than quantities demanded, are employed in several studies reported in this book. One study employs a quadratic programming model for the analysis of the impacts of potential environmental controls related to organo-chlorine insecticide and nitrogen fertilizer usage in agriculture. This nonlinear programming model has 103 producing areas, 10 irrigated crop producing regions, and 10 market regions. The endogenous commodities for which both prices and production are determined are cattle, calves, hogs, milk, oilmeals, wheat, corn, oats, barley, sheep and lambs, chickens and turkeys, eggs, cotton, roughage, feeder calves, and yearlings.

Another study utilizes a separable programming model to analyze inter relationships between the two policy objectives of reducing farm output to increase prices and farmer incomes and reducing gross soil erosion to conserve agricultural land resources. The separable programming model has 105 producing areas, each with 5 land groups of agricultural land, 28 market regions, and 8 demand regions. This nonlinear programming model has endogenous crop and livestock sectors for barley, corn, corn silage, sorghum, sorghum silage, soybeans, wheat, beef cows, feeders, dairy, and hogs.

Hybrid Model

The hybrid model is a model with a mathematical programming component incorporated directly within an econometric simulation model. The combination of these two modeling techniques is aimed at projecting a time path of agricultural adjustment to policy changes. This model's programming component is used to simulate crop production for the state of Iowa. The results obtained from the programming component are aggregated with the production predicted by the econometric simulation component for the remainder of the U.S. agricultural sector. The aggregated results are utilized in the determination of national crop prices.

NOTES ON NOTATION AND PRESENTATION

The studies in the volume are not made by means of a single model. Rather, a series of models are used to analyze varying problems relating to land and water use and conservation, the environment, technology and its producers' market policy, and institutional impacts. Even though some of these models are drawn from the same general methodology, they vary greatly in dimension, specification, and nature. Thus, it is not possible to present a single model and then explain all problem solutions in terms of it specifically. In some cases, parts of the models and their construction are common. In these cases we do not repeat the data, figures, or equations but make references back and forth among chapters. In other cases, the models used in different chapters are of altogether different specifications, and if their nature, application, and results are to be understood, it is necessary to include these extensions as they are made or modified in successive chapters.

In general, a consistent notation is used. However, there are cases in the methodology sections where this is not necessary or useful. For the most part, anyone who knows matrix algebra and calculus will understand the notation. Where it is most convenient to do so, a notation is used which simply represents a set of vectors or matrices multiplied by each other. In other cases, where the detail is useful, the same general relationship may be expressed as a set of equations in which multiplications are denoted by the summation of products of the elements of vectors and matrices. When nonlinear models are involved it is necessary to revert to calculus notations.

In the chapters which follow, we first describe the general profile or societal problem involved. We next *summarize* the model (or the modification of a model) used to solve it. We then explain the results of the model applied to the problem. Finally, at the end of each chapter, we provide a narrative summary that does not make use of any mathematics. Advanced mathematics is not necessary to understand the problem, the results, or the summary. The entire book is organized in this way.

REFERENCES

Alt, K., et al. 1981. Implications of Land, Water, and Energy Resource Policies on Agricultural Production. Staff Report AGESS810513. Economics and Statistics Service, USDA, Washington, D.C.

Council for Agricultural and Science and Technology. 1981. Preserving Agricultural Land: Issues and Policy Alternatives. Report 90. CAST, Ames, Iowa.

___. 1982. Soil Erosion: Its Agricultural, Environmental, and Socioeconomic Implications. Report 92. CAST, Ames, Iowa.

Heady, E. O. 1983. Models for Policy Purposes. The CARD Example. *European Review of Agricultural Economics* 10(1):1-15.

Langley, J., W. Huang, and E. O. Heady. 1981. A Recursive, Adaptive Hybrid Model for the Analysis of National Policies. CARD Report 100. Center for Agricultural and Rural Development, Iowa State University, Ames.

Schatzer, R. J., and E. O. Heady. 1982. National Results for Demand and Supply Equilibrium for Some U.S. Crops in 2000: Theory and Application of Tatonnement Modeling. CARD Report 106. Center for Agricultural and Rural Development, Iowa State University, Ames.

United States Department of Agriculture. 1980. 1977 National Resource Inventories. Soil Conservation Service, Washington, D.C.

___. 1981a. Agricultural Food Policy Review: Perspective for the 1980s. AFPR-4. Economics and Statistics Service, Washington, D.C.

___. 1981b. 1980 Appraisal Part I. Soil, Water, and Related Resources in the United States: Status, Condition, and Trends. U.S. Government Printing Office, Washington, D.C.

___. 1981c. A Time to Choose Summary Report on the Structure of Agriculture. U.S. Government Printing Office, Washington, D.C.

Wisner, R. N., and M. S. Denbaly. 1983. World Food Trade and U.S. Agriculture, 1960-1982. World Food Institute, Iowa State University, Ames.

SECTION I. INTERREGIONAL ASSESSMENT OF SOIL CONSERVANCY PROGRAMS FOR AGRICULTURE

TWO interregional competition studies are reviewed in this section. They focus on quantifying the consequences of reducing soil erosion losses from cultivated lands. The objectives of the first study were to formulate and specify a programming model of the U.S. agricultural sector to answer such questions as: Does U.S. agriculture have the capacity to meet projected domestic and foreign demands for the year 2000 at a reasonable cost to consumers while simultaneously reducing soil losses? What are the interregional consequences for U.S. agriculture if farmers are required to comply with erosion control measures? How are resource values and returns affected? What patterns of soil loss occur? These questions are addressed in the context of a national soil erosion control effort.

The second study in this section addresses the same questions but in a different context. Instead of a national soil erosion control program, as in the first study, only one state, Iowa, is required to comply with a soil erosion control effort. Attention is given in particular to sacrifices Iowa farmers might have to make relative to farmers elsewhere in the country if Iowa farmers alone had to adhere to a mandatory soil erosion control program. Iowa, in fact, has a Soil Conservancy Act, which it could use to cause farmers to attain these goals (if the state had the money to enforce the program).

The linear programming models employed in both studies are developed from the same data base. The formulation of each model is explained in its respective chapter.

The evaluation of a nationwide soil erosion program with the interregional model indicates that agriculture can reduce soil losses moderately with only small increases in the farm level prices of agricultural commodities. More stringent restrictions, however, result in marked increases in commodity prices, especially when in combination with high export demands. Total soil loss can be significantly reduced through a shift to conservation practices and reduced tillage. Some changes in crops grown are indicated, as the less erosive crops are substituted for the more erosive crops, especially the silages.

Interregionally, a nationwide erosion control effort results in pronounced shifts in production patterns in the South Central and South Atlantic regions of U.S. agriculture and sacrifices in income for these regions. The regions in the West, where runoff is lower, gain in production of agricultural products. In all regions, the more erosive crops are incorporated into rotations that reduce the intensity of rowcropping and are grown on lands less subject to erosion.

The assessment of an effort by a single state, Iowa, to change land use and improve environmental conditions results in the conclusion that incomes and land values will be reduced significantly in Iowa relative to the rest of the country. The improved environment in Iowa resulting from the restriction on soil erosion does not, however, impose significant burdens on consumers in the form of higher prices for agricultural commodities. The small impact on consumers is the result of interregional adjustments in production patterns as areas outside of Iowa gain from the reduced competitiveness of Iowa within the U.S. agricultural sector.

CHAPTER 1. INTERREGIONAL COMPETITION MODELING OF A NATIONAL SOIL CONSERVANCY PROGRAM

by Ken Nicol and Earl O. Heady

THIS STUDY evaluates impacts on the agricultural sector from possible imposed limits on sheet and rill erosion of cultivated lands. The main questions addressed are: Does agriculture have sufficient production capacity to meet domestic and export demands and also contribute to improvement of the environment through reducing the quantity of sediment discharged into the nation's waterways? If so, how far can environmental improvement through reduction of soil loss from agriculture be carried while food demands are met at reasonable real prices? What long-run interregional changes in crop and livestock production and land use are favored economically as different levels of environmental improvement through soil loss controls are attained? The results from the study provide insight into potential changes in the cropping and land-use patterns that are economically and technologically feasible given the imposed restraints. Price-related effects analyzed include changes in food costs to consumers and changes in the values of farm resources.

Answers to these types of questions require a model that can generate detailed empirical results at both national, regional, and soil-group levels. National detail is required so that market impacts on prices can be measured, supplies can be equated with demand, and interdependence among regions can be established. Detail by region and soil group is needed so that the flexibilities of or restraints on production and resource use can be measured and impacts can be expressed at the local level. In meeting national food needs, land in one region is a substitute for water in another region, capital in the form of fertilizer used on level land is a substitute for crops on hilly land in another region, and alternative crops are substitutes as feed inputs for livestock within a given region or among regions. This type of detail can be provided by a linear programming model incorporating relevant production possibilities differentiated to reflect their regional technologies.

The formulation of a minimizing linear program in matrix notation is:

$$\begin{aligned} &\min C'X && (1.1)\\ &\text{subject to } A_1X \geq D\\ &\qquad\qquad\;\; A_2X \leq R \end{aligned}$$

where C = a vector of costs; X = the vector of the activities in the model; D = the demands to be met; R = the resources available; A_1 = a matrix of the interaction coefficients between X and D; and A_2 = a matrix of the interaction coefficients between X and R. This formulation is consistent with the regional structure of agriculture if the vector D represents the regional and national demands for the commodities to be met by the system and R represents a vector of regional and national resource availabilities for use in satisfying the demands. The activities in vector X represent production and transformation activities by region and the transportation alternatives connecting the regions in the model. These restraints and activities are defined in the next section. The A matrices include the activity interaction coefficients with the resources or demands. The interactions will be delineated and the quantification procedures outlined in the section following the model formulation.

In this model, the objective function (1.1) minimizes the cost of producing and transporting all endogenous crops. Simultaneously, a long-run equilibrium prevails in the sense that all resources (except land and water) receive their market prices. Returns to land and water are determined endogenously.

The demand restraints represent markets in the model. The resources are obtained in a market, and the production (supplies) and demands interact in markets, including those for intermediate goods such as the feed grains and feeders. Other restraints are used to control patterns of resource use or reflect institutional restraints. Restraints are used where the use or production of one commodity requires a nonmarket, but fixed, interaction with another endogenous commodity or with factors not directly controlled by the model interactions. These restraints generally are in the form of bounds restricting activities to some level not regulated by the resource or product market systems. (An endogenous commodity is a commodity whose production is determined when the model is solved. The model also incorporates exogenous commodities, which are commodities whose production level is predetermined, i.e., not determined when the model is solved.)

DELINEATION OF THE MODEL

Because of the wide variation in climate, soil, and farm practices within the American agricultural sector, a model to determine the long-run impacts of proposed environmental policies should reflect possible regional adjustments that may occur when a change in policy affects regional comparative advantages. The model must incorporate the long-run forces of interregional competition created by the regional variation in natural endowments and production possibilities. The interregional competition model employed in this study is based on sets of regions consistent with the available resources, possible production techniques, and the interactions to be examined. Within the relevant regions, restraints are imposed on the interaction between resource availability and use and commodity requirements. This section gives an outline of the structure of the model and an explanation of the determination of its coefficients (Nicol and Heady, 1974).

Structure of the Model

The model utilizes three sets of operational regions in delineating the interactions among resources, production, and demand. Restraints are defined in these regions based on the availability of land and water resources and the demands for the crop and livestock commodities. Activities define, by region and soil-resource groups, the possibilities for crop and livestock production to transform these resources into agricultural commodities to meet the demands.

Resource availability restraints define, by each quality group, the number of acres of land that are available in each producing region for cropland production, including cropland, hay, and pasture. This land base is adjusted for the requirements of crops whose regional distribution is not determined within the model. Additional land resource restraints include the land available for nonrotation hay and pasture and for forest land grazed by region. Also included are restraints on the availability of water for agricultural uses and on the availability of nitrogen fertilizer.

The production activities utilize the resources to produce the crop and livestock products for both intermediate and final uses. Included are the crop and livestock production alternatives related to environmental considerations, interregional transportation of the transportable commodities, and product transformation activities.

The demand restraints define on a regional basis the final demands for the commodities. These commodity demands are determined by the domestic consumption levels for the commodities, the domestic requirement for livestock production outside the model, and the level of exports.

Regional Delineations in the Model

Five separate sets of regions are employed to define the model. The first set is the regions within which the resource data base is defined; the second, the regions within which the production activities are defined (to be referred to as producing areas); the third, the regions detailing both water availability and water transfer possibilities; the fourth, the regions within which the commodity markets are defined; and the fifth, the regions into which the results are aggregated for reporting.

The data regions. These regions represent many sets of political and geographic areas within which data are tabulated by various collecting agencies. They include the counties and states of the continental United States within which census and commodity production data are tabulated. An additional set of regions included in this group is the county approximations of the major land resource areas, as used for data collection by the Soil Conservation Service, USDA (USDA, 1967 and 1971). These regions divide the land in the continental United States into 164 areas based on soil type and management characteristics. It is from these regions that the data used in calculating the soil loss by alternative cropping activities, tillage methods, and conservation practices are developed.

Sets of weights based on relevant data relationships are used to transfer data from these data regions into the model's producing areas where the data are used directly in the model or in combination with other data to generate coefficients to be used in the model.

The producing areas. The producing areas are the regions in which crop and livestock production activities, land by quality class, pasture, and nitrogen balance restraints are defined. Figure 1.1 displays the 223 producing areas defined for the model. These areas are based on county approximations of the Water Resource Council's subareas (Water Resources Council, 1970) modified to be consistent with the water supply regions and the market regions. Each producing area is an aggregation of contiguous counties contained in a watershed draining to a common waterway. Each producing area contains 9 land groups.

The water supply regions. Fifty-one water supply regions define the areas in the seventeen western states where water supplies are incorporated in the model (Figure 1.2). These regions are aggregations of contiguous producing areas within which a water supply can be said to exist. The subdivisions of the 18 major river basins of the Water Resources Council are the basis for these regions (Water Resources Council, 1970). Each of the 51 water supply regions contains 18 land groups −9 potentially for irrigated crops and 9 potentially for dryland crops.

The market regions. Contiguous producing areas are aggregated into 30 market regions for the model (Figure 1.3). Defined within these regions are the market balance restraints equating output of agricultural products with their requirements. The market regions each have a city that serves as their market center in the model's national transportation network.

The reporting regions. These regions are aggregations of the market regions such that regional similarities in agricultural production possibilities are maintained. The 7 reporting regions neither overaggregate the production impacts nor create a reporting system that is completely overpowered by numbers. (The regions are North and South Atlantic, North and South Central, Great Plains, North and South West.)

The Land Base

The land base represents the major constraint on the productive capacity of U.S. agriculture in the model. The number of acres of dryland and irrigated cropland for use by the endogenous crops, nonrotation hays, and nonrotation pastures are determined by aggregating the county acreages from the Conservation Needs Inventory (CNI) (USDA, 1971). The CNI reports the acreage of privately owned land by use and by agricultural soil capability class as determined from a 2 percent sample of all private lands in the nation. The soils classification scheme

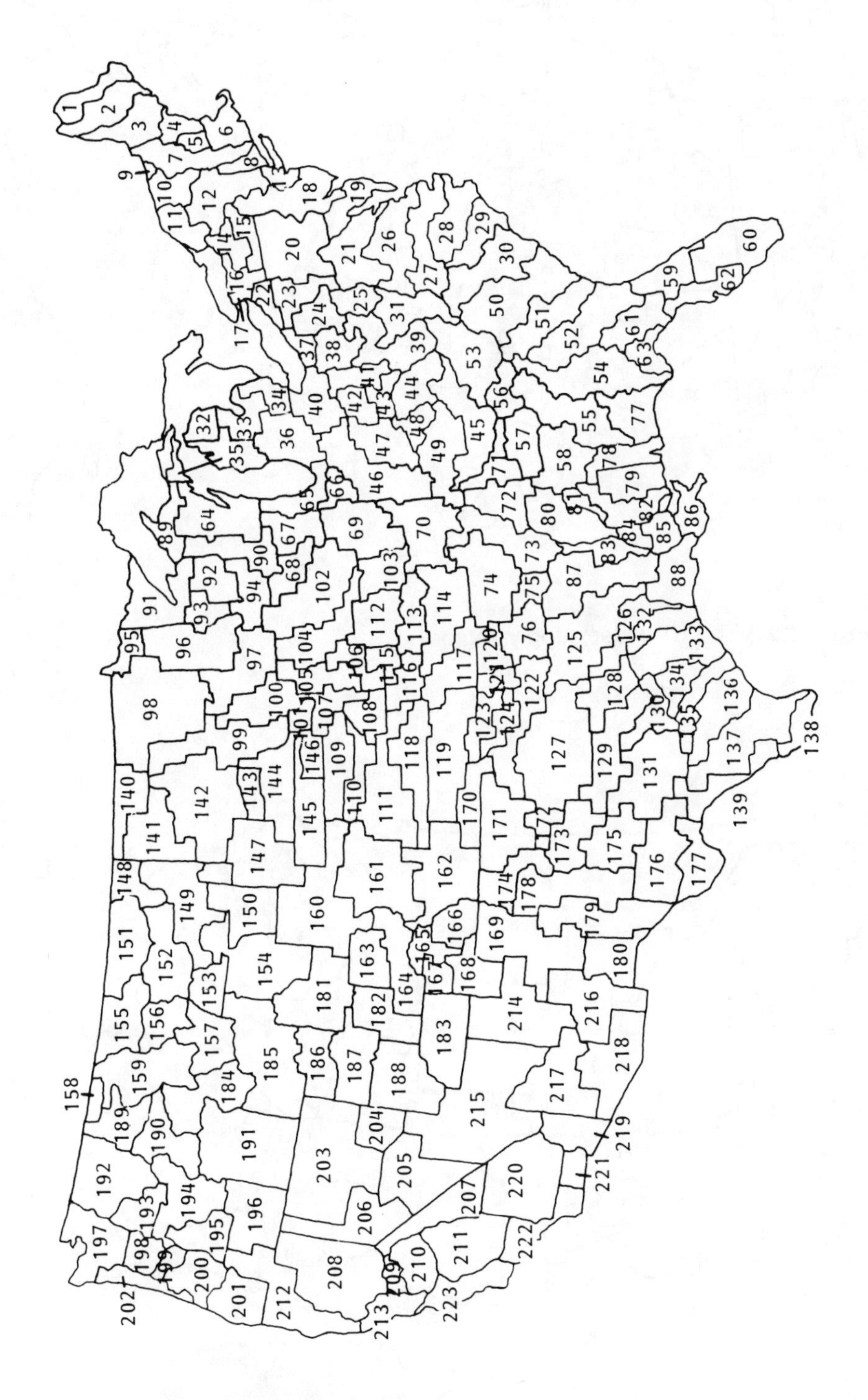

Figure 1.1. The 223 producing areas.

Figure 1.2. The 51 water supply regions.

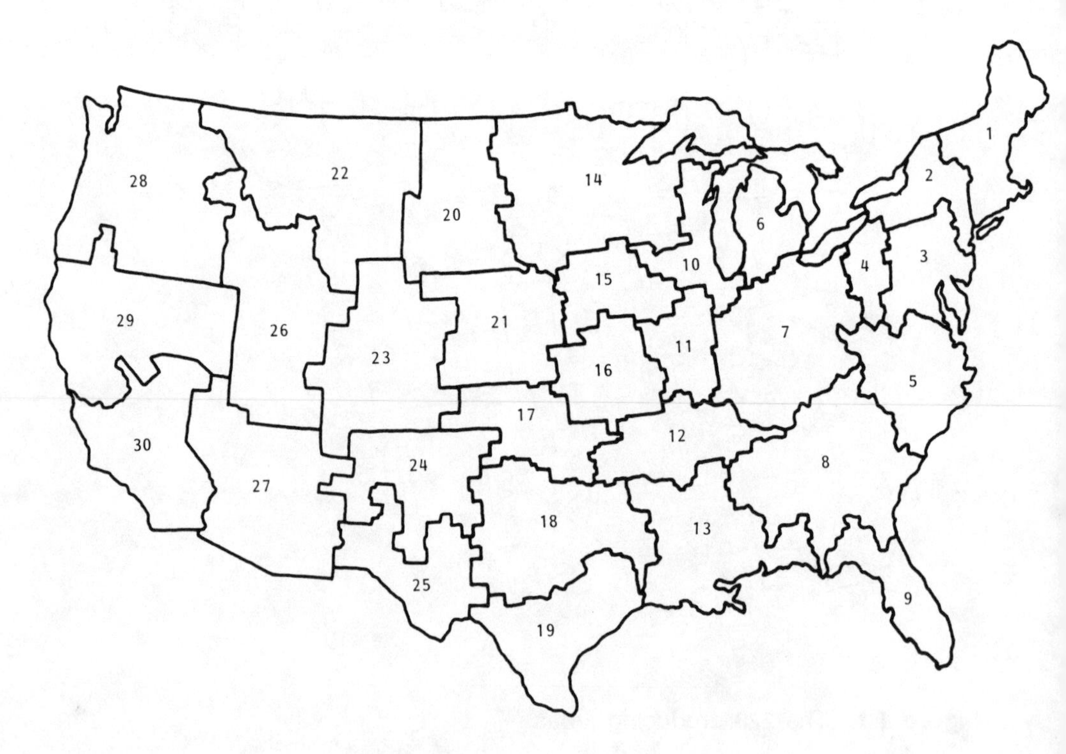

Figure 1.3. The 30 market regions.

of the Soil Conservation Service (SCS), USDA, for the CNI places all soils within eight capability classes. The risks of soil damage or limitations in use become progressively greater from Class I to Class VIII. In general, the first four land capability classes are arable soils capable of producing crops without deterioration over a long period, if under proper treatment. These soils may be used for pasture, range, forest, and woodland. Soils in land capability classes V, VI, and VII are primarily nonarable soils suited mainly for grasses or trees. Soils in land capability class VIII are not suited for crops, grass, or trees. Class VIII soils include rock outcrops, canyons, bluffs, and river washlands. Briefly, the eight capability land classes are defined as follows:

Class I – Soils with few limitations restricting their use.
Class II – Soils with some limitations that reduce the choice of plants or require moderate conservation practices.
Class III – Soils with severe limitations that reduce the choice of plants, require special conservation practices, or both.
Class IV – Soils with very severe limitations that restrict the choice of plants, require very careful management, or both.
Class V – Soils with little or no erosion problems, but which have other limitations that are impractical to remove and that restrict their use largely to pasture, range, woodland, or wildlife food and cover.
Class VI – Soils with severe limitations that make them generally unsuitable for cultivation and that limit their use largely to pasture, range, woodland, or wildlife food and cover.
Class VII – Soils with very severe limitations that make them unsuitable for cultivation and that limit their use largely to grazing, woodland, or wildlife.
Class VIII – Soils and land forms with limitations that preclude their use for commercial plant production without major reclamation and that restrict their use to recreation, wildlife, water supply, or aesthetic purposes.

Additionally, the CNI subdivides soil capability classes II through VIII into subclasses to reflect the most severe hazard preventing the land from being available for unrestricted use. For capability class I there is no subclass because it includes only those soils with few or no limitations. The capability subclass is shown by a letter after the land capability class. These letters stand for the principal kind of problem or limitation applicable to the land subclass.

e – Deep soils with favorable physical conditions throughout, occurring on erosive slopes. Soils with favorable upper layers but rock or other restrictive layers within 24-36 inches. Soils with very tight or hard subsoils with moderately heavy subsoils that are of slow or moderately slow permeability.

w – Soils with moderately heavy subsoils occurring on flat or depressed areas. Soils with tight subsoils with very slow permeability. Organic soils. Bottomland soils with claylike surface soils and very slowly permeable subsoils. Wet soils that are sandy.

s – Well-drained soils, favorable throughout but having sand or rock layers within 30 inches. Excessively drained soils, sandy throughout. Stony soil of variable depth and texture-clearance of stones not practical.

Limitations imposed by erosion, excess water, shallow soils, stones, low moisture-holding capacity, salinity, or alkalinity can be modified or partially overcome and take precedence over climate in determining subclasses. The dominant kind of limitation or hazard to the use of the land determines the assignment of capability units to the e, w, and s subclasses. Capability units, which have no limitations other than climate, are assigned to a c subclass.

Besides reporting the land by agricultural capability, the CNI also reports the acreage of privately owned land by agricultural use by county. For the model's cultivatable land base the CNI county acreages are aggregated, for dryland and irrigated uses, to the 223 producing areas by the twenty-nine capability-class subclasses. The twenty-nine capability classes are aggregated to give 9 land groups that exhibit a range in erosion hazard, yield potential, and farming alternatives (Table 1.1). The land base used for the dryland and irrigated crops is the sum of the acres in the component land classes of the CNI designated as being used for row crops, close-grown crops, summer-fallow, rotation hay and pasture, temporarily idled cropland, and land used for fruits and vegetables.

Projected increases in irrigated lands in the western United States are added to the irrigated acreages in each of the relevant producing areas. Only those irrigation projects that have been approved for construction before 1980 are

Table 1.1. Dryland and irrigated acreages aggregated to nine land groups by class and subclass to create the land base for endogenous crops

Land Group	Inventory Class-subclass[a]	Dryland	Irrigated
		------(thousand acres)-------	
1	I	23,458	5,632
2	IIe	76,672	7,257
3	IIs, IIc, IIw	73,748	4,796
4	IIIe	65,598	3,648
5	IIIs, IIIc, IIIw	45,838	4,120
6	IVe	29,034	1,410
7	IVs, IVc, IVw	10,738	1,168
8	all of V	305	14
9	all of VI, VII, VIII	12,829	287
		338,220	28,332

[a]Inventory classes and subclasses are as defined by the Soil Conservation Service (SCS) for the Conservation Needs Inventory (CNI) (U.S. Department of Agriculture, 1971).

considered. The projected increase in acreage is weighted to the relevant land groups based on the proportion of the irrigated acreage presently in each group as determined from the CNI. A corresponding number of acres is removed from dry cropland and pasture in proportion to their acreages in the area as indicated in the CNI. The total dryland and irrigated acreages in the 9 land groups available for the endogenous crops is shown in Table 1.1.

The noncultivated land base for hays and pastures is divided into 3 land-use categories and is based on acreages from the CNI. Dryland nonrotation pasture and rangeland are combined to give an upper bound for the improved or managed pasture activities by producing area. Irrigated, improved pasture acreages are obtained from the irrigated nonrotation pasture and rangeland categories of the CNI data. The dryland and irrigated noncropland hay lands of the CNI are aggregated to provide upper bounds for the dryland and irrigated permanent hay activities in the model. These acres represent wild hay and other hayland, which is continuously harvested except for infrequent interruptions to reestablish or improve the stand. The third permanent-use category of the noncultivated land base is the grazed forestland category from the CNI. This land represents the relatively low-yielding woodlands pasture on farms, as well as the large tracts of forested lands under private control.

The Water Sector

The water supply in each water supply region is a function of both the total reservoir storage and the mean annual runoff in the region. The total storage capacities of reservoirs in each of the water supply regions are determined by adding the active conservation and joint-use capacities for storage dams in the region obtained from the Bureau of Reclamation, the Army Corps of Engineers, and a survey of reservoirs in the United States. (See Nicol and Heady, 1975, for details about sources.) The mean annual runoffs were determined from the Water Resources Council (1968). Then using the relationship between storage and mean annual flow developed by Lof and Hardison (1966), the net water supply as a proportion of the mean annual runoff is determined.

In using the work by Lof and Hardison, it is assumed that all water supply regions in a given river basin have the same relationship between the gross water supply and total reservoir storage. Gross water supplies are first calculated for all water supply regions. Then the gross water supplies are adjusted for reservoir evaporation, based on the work by Lof and Hardison, to give a net water supply in each of the water supply regions.

Mining of underground water supplies is not permitted in any of the water supply regions. Those underground sources that are replenishable remove surface runoff, and only the system of distribution (pumps versus diversion canals) differs. Also, since the mean runoff includes some unknown amount of water leaving the surface runoff channels and entering underground streams, inclusion of certain underground water supplies would increase the net water supply above its true amount as a result of double counting. At the same time, some of the water returning from canal losses and farm wastes enters the underground streams and later emerges as surface runoff. Thus, more double counting would result if these return flows are added to the water supply.

The price presently paid by farmers for water in each water supply region was determined using a weighted average of present water costs in Bureau of Reclamation irrigation projects (U.S. Department of the Interior, 1970). For regions in which Bureau of Reclamation data were not available, the water price in the most immediate upstream region was used. These estimated water prices were increased to account for farm waste and deep percolation to give water prices based on cost per acre foot of water consumed. No correction was required for canal losses since the deliveries were measured at the farm.

Water transfer activities are defined to allow water in upstream regions to flow to downstream water supply regions. Each of these activities is bounded at a maximum of 70.0 percent of the upstream water supply. Since losses occur from evaporation, removal by natural vegetation, and some deep percolation, this restraint prevents downstream movement of water with 100.0 percent efficiency. The costs associated with these natural flow transfers are set so that the upstream water price plus the transfer cost is greater than the price of water in the receiving region. For some of these activities the cost would be zero if water in the upstream region were priced higher than water in the downstream region. This procedure is included to force regions to use locally available water before using the upstream water. This guarantees that all water used in the region will be priced at least equal to local water.

Existing interbasin transfers are simulated by transfer activities for those projects. Because the facilities are fixed and since present water prices reflect their variable cost, no cost is directly attached to the transfer. An upper bound set at the projected capacity of the project to transfer water in the year 2000 controls the level of flow.

Unbounded water-desalting activities are defined for all sea coast water supply regions to allow for augmentation of the water supply. The price of $100.00 per acre foot placed on these activities approximates the best available estimates of the cost of large-scale desalting schemes under present technologies (Howe and Easter, 1971).

Per capita water consumption estimates by region for recreation, municipal and industrial uses, and thermal electric power are multiplied by the projected population and the total subtracted from the available water supply in each water supply region. (See Nicol and Heady, 1975, for detailed sources.)

The Crop Sector

Activities representing the production of the endogenous crops are defined by land group in each producing area of the model. These activities represent crop management systems incorporating a rotation of one to four crops, covering one to eight years, with alternative conservation treatments and alternative tillage practices. The crop rotations defined for each producing area and land group are selected from 330 unique rotations developed from the Soil Conservation Service (SCS) questionnaire (Nicol and Heady, 1975). The rotations in each producing area are selected to give a range of production alternatives consistent with production practices and patterns projected for the year 2000. Using crop rotation activities instead of individual crop activities allows for the determination of interrelationships such as the fertilizer carryover for crops following legume crops and the characteristics of certain crops to provide large amounts of residue to help reduce erosion. The selected rotations for each land group are combined with one of four conservation treatments: straight-row farming, contouring, strip farming, or terraces. Conservation treatments are defined on the land groups according to the recommendations given in the SCS questionnaire. The crop management system is completed by specifying one of the three tillage practices: conventional tillage with residue left, conventional tillage with residue removed, or reduced tillage. Soil loss, crop yields, fertilizer use, costs, and water use coefficients are calculated for each of the crop management systems.

Determination of the soil-loss levels. Gross soil loss, as calculated in the cropping activities, is the average tons of soil leaving the field annually. This measurement of soil loss does not represent the amount reaching waterways, since some soil particles settle out or are diverted as the runoff passes through grassed areas or onto flatter terrain. Two separate procedures were used to determine the gross soil loss per acre: one procedure for areas east of the Rocky Mountains and another procedure for the Rocky Mountains. For the areas east of the Rocky Mountains, the Universal Soil-Loss Equation (USLE), as described by Wischmeier and Smith (1965), is used to develop the gross soil coefficients. The soil-loss equation is presented by:

$$A = R \cdot K \cdot L \cdot S \cdot C \cdot P \tag{1.2}$$

where A is the average annual per acre soil loss; R is a rainfall erosive factor based on the local area's rainfall patterns; K is a soil erodibility factor for the specified soil determined from its erosion under continuous fallow on a nine percent slope, 72.6 feet long; S is the slope gradient factor relative to a nine percent slope; L is the slope length factor relative to a 72.6 foot slope length; C is the crop management factor which relates to a particular crop rotation and tillage practice; and P is the erosion control practice factor which relates to the conservation practice. Further detail on the factors and on the computational procedures used to calculate them is available from Wischmeier and Smith (1965) and from the Soil Conservation Service (1972). For the areas east of the Rocky Mountains, the above variables are defined as the dominant value existing on each soil class and subclass in the area of reporting. The soil loss is then computed by the land resource area of the SCS questionnaire for each feasible combination of crop rotation, conservation practice, tillage method, and soil group defined from the SCS questionnaire.

The soil loss determined for the twenty-nine soil classes and subclasses is aggregated using weighting functions determined from the CNI data to get soil-loss coefficients for each of the 9 soil groups of each producing area. The determination of the soil loss by cropping management system is weighted to the producing area from the SCS data regions as follows:

$$S_{ijm} = \sum_k SL_{ijk} A_{jkm} / A_{jm} \qquad (1.3)$$

i = 1,..., the number of crop management systems defined in the producing area;
j = 1,..., 9 for the land classes;
k = 1,..., 165 for the parts of the SCS data regions;
m = 1,..., 223 for the producing area.

where S_{ijm} is the soil loss for crop management system i on soil group j in producing area m; SL_{ijk} is the soil loss from crop management system i on soil group j consistent with SCS data region k; A_{jkm} is the acres of tillable soil group j in the part of SCS data region k in producing area m; and A_{jm} is the total tillable acres of soil group j in producing area m.

For those agricultural lands in the mountain valleys and on the West Coast, the data required for the soil-loss equation have not been completely developed and an alternative procedure is used to estimate the soil loss from these lands. The SCS questionnaire asked for crop management systems consistent with the production possibilities in the SCS data areas. The SCS personnel estimated the tons of soil loss by crop management system on each land class and subclass defined in the SCS data area. These estimates are, for purposes of this model, treated as if they were developed from the same procedure as the estimates in the eastern area. Each of the activities representing the production of irrigated crops is considered to have a soil loss similar to the corresponding dryland activities.

Development of the crop yield coefficients. A unique yield is determined for all irrigated and dryland cropping activities as a function of the producing area, soil quality group, the crop rotation, the conservation practice, and the tillage method. The determination of the yield begins with functions projecting crop yields by state. These state functions are modifications of the functions developed by Stoecker (1974). These functions are weighted to producing areas and then yields are adjusted for crop rotation, land quality group, and conservation and tillage practices of each activity. For each crop the function is of the form:

$$Y(t) = Y_o(t) + (1 - .8^{X(t)}) * PF(t) \qquad (1.4)$$

where $Y(t)$ is the estimated average per acre yield of the crop in year t; $Y_o(t)$ is the estimated average per acre yield on unfertilized land in year t, developed from a linear trend function; $X(t)$ is the number of units of fertilizer applied to each acre of the crop in year t; $PF(t)$ is the proportion of the acreage of the crop receiving fertilizer in year t, developed from a linear trend of the proportion of the crop acres receiving fertilizer; and t is years after 1949. The X(t) above represents:

$$X(t) = PO(t)*(\ln(Px/Pc) - \ln A - (\ln(- \ln .8)))/\ln .8 \quad (1.5)$$

where 1n is the natural log of base e; Px is the weighted price of a unit of fertilizer; Pc is the price of a unit crop c; and PO(t) is the proportion of the optimum rate of fertilizer applied in year (t), developed from a linear trend of the proportion of the optimum rates applied.

The next step in determining the yields for a cropping activity is to adjust for land class, conservation practice, and tillage method. The data obtained in the SCS questionnaire includes a set of ratios giving the land class yields of each crop relative to the most productive land class of the area. These ratios initially are weighted to the 9 land groups and are adjusted such that land group 1 has a relative yield value of 1.00. The acreage weights used are the acres of the respective crop categories, row crops, close-grown crops, and rotation hay and pasture, from the CNI. The producing area yield is determined as a weighted average yield over the land groups in the producing area. Using the relative yield indices, the weighted average function is expressed as:

$$\begin{aligned} Y_{ij} &= W_{ij1}Y_{ij1} + W_{ij2}R_{ij2}Y_{ij1} + \cdots + W_{ij9}R_{ij9}Y_{ij1} \quad (1.6) \\ &= (\textstyle\sum_k W_{ijk}R_{ijk})Y_{ij1} \end{aligned}$$

i = 1, 2, ..., 223 for the producing areas;
j = 1, 2, ..., 30 for the dryland and irrigated crops;
k = 1, 2, ..., 9 for the land groups.

where Y_{ij} is the average yield of crop j in producing area i; W_{ijk} is the weight of acres in producing area i which are on land group k for crop j; R_{ijk} is the relative yield factor for crop j on land group k in producing area i with land group 1 = 1.00; and Y_{ij1} is the yield of crop j on land group 1 in producing area i. The above equation can be transposed and solved for Y_{ij1}, and each of the other land group yields can be subsequently determined from the land group 1 yield using the relative yield indexes.

The conservation and tillage yield ratios, obtained from the SCS questionnaire, are used equally on each land class to adjust the yields for both conservation and tillage effects. The national average ratio is used as a proxy for the adjustment ratio in a producing area if the area's data are missing. These adjustments complete the calculation of the crop yields as determined from the response function of the area, the land class, the rotation, the conservation practice, and the tillage method.

In addition to estimating the yield of the crop harvested, the yield of available aftermath pasture must be estimated. Pasture aftermath is the pasture available for livestock after harvesting the crop, when animals are allowed on the field to graze the aftermath and fence rows. Pasture aftermath yields are adjusted from the work of Jennings (1955). Jennings estimated the output of aftermath pasture in acres of cropland pasture equivalent for each of the forty-eight contiguous states. This output is divided by the total acres of cropland and hayland in the state to give an average yield per acre of dryland and irrigated land cropped. Each county in the state was assumed to have a yield of aftermath pasture, and the producing area yield is obtained as a weighted average of all the counties in each producing area. These yields are included as roughage production with the annual crop activities, except for soybeans and cotton. For the hay crops, these yields are added to the roughage production. The aftermath pasture yields are affected by the tillage method used.

Fertilizer used coefficients for the crops. The fertilizer use coefficients are developed from the crop yield functions independently of the land quality group, the conservation practice, and the tillage method. The yield functions adapted from Stoecker (1974) provide the basis for determining the level of nitrogen fertilizer application. The nitrogen coefficient in the crop activity is determined by taking the optimum level of fertilizer use as determined from the function and subtracting the amount provided by the legumes, if any, in the rotation. The nitrogen contribution of the legume crop is determined using functions which relate nitrogen fertilizer equivalent carryover of the legume crop for the following two years as a function of the yield of the legume (Nicol and Heady, 1975). A similar functional relationship has been developed for nitrogen carryover from soybeans. Shrader and Voss (1972) showed that soybeans provide a carryover to the next year of approximately one pound of nitrogen equivalent per bushel of soybean yield.

The nitrogen for the cropping activities is obtained through the purchase of commercial nitrogen fertilizer or the use of livestock manures. The nonnitrogen fertilizer required to satisfy the calculated optimum application rate is assumed to be purchased, and the cost of this fertilizer is included in the variable costs for the cropping activities.

Development of the crop production costs. The production costs incorporated in the cropping activities are adapted from work of Eyvindson (1965). Eyvindson developed a set of crop budgets for the crops of barley, corn, corn silage, cotton, tame hay, wild hay, oats, sorghum, sorghum silage, soybeans, and wheat. In areas where irrigation is relevant, he developed both dryland and irrigated budgets. The procedure used was one of budgeting each crop based on the most common production technique in the area. This entailed determining machinery sequences for each crop, machinery size, average life of the machines, repairs needed, and the acres covered with the machines. These data are combined with the costs of machinery and supporting inputs to provide the cost and labor coefficients for each of the crop budgets. The budgets were developed to include all costs except return to land or any fixed cost associated with the land. Labor costs were adjusted to account for changes in technology consistent with a continuation of trend levels. Total variable costs for each crop are projected to the year 2000, assuming constant per unit costs.

Adjustments for conservation practice and tillage method are determined from the SCS questionnaire. Straight-row cropping is used as the base for comparisons made in machinery and labor efficiency for contouring, strip cropping, and terracing. Similarly, adjustments are made for the tillage practices with conventional tillage with no residue left on the field as the base. The variations include conventional tillage with residue left on the field as the base. The variations include conventional tillage with residue left on the field and reduced tillage.

A further adjustment is made for the reduced tillage operations to reflect the tradeoff between tillage operations and the use of herbicides for weed control. In areas which are not moisture deficient, the saving in machinery cost is exactly offset by increased herbicide costs (Nicol and Heady, 1975). In arid areas the adjustment consisted of a $3.00 increase in herbicide costs for each $1.00 reduction in nonherbicide costs (Shipley and Osborn, 1973). This is consistent with the extensive farming methods used and the relatively lower machinery cost per acre when compared to the fixed herbicide application cost.

Summer-fallowing is treated as a separate crop in the rotations. A cost adjustment relationship is developed by comparing the crop rotations in the *Selected U.S. Crop Budgets* (Economic Research Service, 1971a and 1971b) that include summer fallow to those rotations that do not have summer-fallow. In this way an estimate of summer-fallow costs is obtained and a ratio of summer-fallow cost to crop cost is developed.

A cost adjustment is made to reflect the terracing costs for those cropping activities defined to include terracing. The SCS questionnaire provides estimates of the construction costs for terraces. The average terracing cost per acre is calculated as:

$$TC_{ij} = .1\,(CC_{ij} + PW_{ij}W_{ij} + PT_{ij}T_{ij})PLT_{ij} \tag{1.7}$$

$i = 1,\ldots, 223$ for the producing areas;
$j = 1,\ldots, 9$ for the land groups.

where TC_{ij} is the per cultivated acre terracing costs on land group j in producing area i; CC_{ij} is the per acre construction cost of terraces on land group j in producing area i; PW_{ij} is the proportion of acres of land group j terraced having grassed waterways for drainage in producing area i; W_{ij} is the cost per terraced acre for grassed waterways consistent with the terraces on land group j in producing area i; PT_{ij} is the proportion of acres of land group j terraced having tiled outlets for drainage in producing area i; T_{ij} is the cost per terraced acre of tiling and drainage consistent with the terraces on land group j in producing area i; PLT_{ij} is the proportion of all land in group j which is feasible to terrace in producing area i; and .1 is the factor to adjust for a ten-year amortized life of the terrace. The total of these costs for each cropping activity is:

$$C_{ijk} = \sum_{m} (M_{ijm} + L_{ijm} + P_{ijm} + F_{ijm} + MS_{ijm})R_{ijm} + TC_{jk} \quad (1.8)$$

i = 1,..., for the number of cropping management systems in the producing area;
j = 1,..., 223 for the producing area;
k = 1,..., 18 for the land groups;
1,..., 9 dryland, and 10,..., 18 irrigated land groups;
m = 1,..., 15 for only those crops in the cropping system.

where C_{ijk} is the cost per acre for crop managment system i in producing area j on land group k; M_{ijm} is the projected per acre machine cost for crop m in cropping system i in producing area j; L_{ijm} is the projected per acre labor cost for crop m in cropping system i in producing area j; P_{ijm} is the projected per acre pesticide cost for crop m in cropping system i in producing area j; F_{ijm} is the projected per acre nonnitrogen fertilizer cost for crop m in cropping system i in producing area j; MS_{ijm} is the projected per acre other costs for crop m in cropping system i in producing area j; R_{ijm} is the rotation weight for crop m in cropping system i in producing area j; and TC_{jk} is the terracing cost per cultivated acre on land group k in producing area j.

Crop water use coefficients. Water use coefficients for each crop activity in the model reflect the net diversion requirements to provide the crop with the amount of water needed for growth, in addition to that provided from precipitation. Withdrawal coefficients are also calculated to indicate the diversion requirements needed to supply the water consumed. Gross delivery requirements in area i for crop j are:

$$GDR_{ij} = \frac{CU_{ij} - EP_i}{(IE_j)\ (CE_i)} \quad (1.9)$$

i = 1,..., 223 for the producing areas;
j = 1,..., 19 for the endogenous crops.

where GDR_{ij} is the gross delivery requirement in acre feet for crop j in producing area i; CU_{ij} is the amount of water required by crop j in producing area i (Nicol and Heady, 1975, for sources); EP_i is the effective precipitation in producing area i representing water available after evaporation and deep percolation are subtracted from the rainfall; IE_j is the irrigation efficiency or the efficiency of the crops in using the water applied. This efficiency is affected by the surface of the land exposed between plants and the ability of the plants to hold the water in the ground for use; and CE_i is the canal efficiency or efficiency of the delivery system between the diversion point and the farm delivery gate. This was calculated for each region from data on Bureau of Reclamation projects.

The net diversion requirements, NDR_{ij}, or the water use coefficients for each of the activities, are calculated as:

$$NDR_{ij} = CIR_{ij} + (1 - RF_i)\ GDR_{ij} - CIR_{ij} \tag{1.10}$$

$i = 1, \ldots, 223$ for the producing areas;
$j = 1, \ldots, 19$ for the endogenous crops.

where GDR_{ij} is the gross delivery requirement in acre feet for crop j in producing area i (Nicol and Heady, 1975, for sources); CIR_{ij} is the crop irrigation requirement of crop j in producing area i; and RF_i is percent of the water not used by the plant which is returned for reuse in the region. This return flow is assumed to be 55 percent for all river basins except the Columbia-North Pacific where 60 percent is used.

The noncropland roughage. The noncropland roughage includes the permanent pasture and hay and the forest lands available for grazing. The land available for such uses has been outlined in conjunction with the definition of the land base. This section will outline the costs and yields associated with these activities. The activities are divided into dryland and irrigated permanent pasture, dryland and irrigated permanent hay, and forest land grazed.

The costs of the permanent pasture activities are developed from the preharvest costs of hay as determined by Eyvindson (1965). The yields are developed as a function of the hay yields in the area. Nonirrigated cropland pasture is assumed to have a yield equal to 75.0 percent of the tame hay yield, if the tame hay yield is less than 4 tons, and 70.0 percent of the tame hay yield if the yield is more than 4 tons. Irrigated yields on cropland pasture are determined by a similar relationship, with the yield being equal to 85.0 percent of the irrigated tame hay yield, if it is less than 4 tons, and 80.0 percent of the tame hay yield if it is greater than 4 tons (Nicol and Heady, 1975). Heady and Mayer estimate that improved pasture yields are equal to 88 percent of cropland production (Heady and Mayer, 1967). This yield is used to give an estimate of production from the acres in the pasture category of the CNI.

Bounded activities are defined for dryland and irrigated nonrotation hay. These activities represent wild hay and other hayland which is continuously harvested except for infrequent interruptions to reestablish or improve the stand. The acreage bounds for these activities are obtained from the hayland category in the CNI. The costs are developed from Eyvindson's permanent hay and wild hay costs. An estimate of wild hay acreage is made from the *1964 Census of Agriculture* (U.S. Department of Commerce, 1967), and this is subtracted from the CNI acreage to give the permanent hay acreage. The yield coefficients are determined from an adjusted fifty-year time trend for dryland and irrigated tame hay and wild hay. The trends were determined from annual crop summaries and the Census of Agriculture. (See Nicol and Heady for details on sources.)

A noncropland roughage activity incorporates the forest and grazed category from the CNI into the model. The roughage production from these acres represents grazing of lands mostly in trees. The yield coefficients are adapted from the relationships developed by Jennings (1955). These give the yield relationship between woodland pasture and cropland pasture by state. It is assumed this relationship holds to the year 2000. The cost for the forestland grazed activity is updated from the grazing rates charged on public lands in the area (Nicol and Heady, 1975).

The exogenous crops. The acreage defined for use by the endogenous crops reflects the requirements for the exogenous crops in 2000. The exogenous crop sector includes broomcorn, buckwheat, cowpeas, dry beans, dry peas, flax, hops, orchards and vineyards, peanuts, potatoes, proso-millet, rice, rye, safflower, sugar cane, sunflowers, sweet potatoes, tobacco, and vegetables. Acreages by state for each crop in 1969 are obtained from the *Census of Agriculture* (U.S. Department of Commerce, 1968) and an average state yield is determined for 1969. Dean (1970) reports yields for the exogenous crops produced in California in 1969 and projected yields for each of the crops in the year 2000. The ratio (yield in 2000/yield in 1969) is determined for each crop in the Dean study. It is assumed that the yields in each state will increase proportionately to those in California, and the above ratios are used to adjust all state yields from 1969 to 2000. Acreage requirements for the year 2000 are computed by dividing the estimated production by the projected yields per acre.

All projections in the exogenous crop sector are made at the state level. The acreage is allocated to the counties within the state on the basis of the proportion of each crop grown in the county as reported in the *1964 Census of Agriculture* (U.S. Department of Commerce, 1967). The acreages of each of the exogenous crops in each producing area are determined by summing the projected acreage of the relevant crops in the producing areas over the subset of counties consistent with the definition of the producing area.

Within each producing area the exogenous crops are grouped into three categories according to their method of cultivation. These categories are row crops, close-grown crops, and orchards and vineyards. Acreages of these three categories are then allocated to different land groups in proportion to the calculated acres of other row crops, close-grown crops, and orchards and vineyards as determined by land class in the CNI. This same procedure is used for both dryland and irrigated acreages. If the projected acreage requirement for the exogenous crops is greater than the acreage available in the land group, the excess acres are allocated either to the land group next closest in erosion-hazard characteristics or to the same land group in an adjoining producing area, depending on the agronomic characteristics of the land groups, producing areas, and cropping pastures required to produce the exogenous crops.

Water allocation for the exogenous crop sector is determined directly from water use coefficients developed similarly to those for the endogenous crops. The developed per acre water use rates are applied to the acres of each of the exogenous crops. The water use for all exogenous crops is aggregated to become a fixed water requirement, which is subtracted from each region's water supply to leave the quantity available for endogenously allocated uses.

The use of nitrogen by the exogenous crops represents a significant demand for nitrogen, especially in the Gulf and West Coast areas. The amounts of nitrogen required by the specific crops are adapted from the work of Ibach and Adams (1967). The quantity used per acre for each of the exogenous crops is multiplied by the acres calculated in the region. The assumption is made that by the year 2000 the average application rate for all acres will be equal to the application rate on the acres fertilized in the Ibach and Adams data.

The Livestock Sector

The livestock sector in the model includes dairy, pork, beef cows, beef feeding, broilers, turkeys, eggs, sheep and lambs, and a general category for "other animals" such as horses, mules, ducks, geese, and zoo animals. Production coefficients are required for all categories but cost data are needed only for the endogenous livestock enterprises. The endogenous livestock include hogs, beef cows, beef feeding, and dairy. Livestock production activities for these livestock are defined for each of the 223 producing areas in the model. The budgets for these activities are developed using a procedure similar to the crop budgets and are based on data adapted from Eyvindson (1965).

Each livestock type by producing area has a set of activities with commodity-differentiated rations. All rations are adapted from the nutrient requirements specified by the National Academy of Sciences (1968, 1970, 1971). The rations are formulated to provide alternative levels of substitution between grains, between roughages and grains, and between roughages given a grain component. These rations reflect research-based recommendations that approximate an optimal level of feeding efficiency and are adjusted in order to account for the "inefficiency" of actual production. By providing alternative rations within the set of activities for each livestock type, linear combinations of these activities provide the model with a larger number of possible rations. The rations for the exogenous livestock are predetermined and are used to create a predetermined demand for the commodities.

In producing areas enclosed within a water supply region, coefficients for water requirements are developed by livestock type.

Livestock wastes historically have been a source of plant nutrients. All livestock considered in the model produce nitrogen wastes which may be utilized in the cropping sector as a fertilizer nitrogen source (Nicol and Heady, 1975).

The Demand Sector

Demand restraints defined in the model at the market region level require production of the commodities consistent with domestic food and fiber, export, and intermediate feed requirements estimated for the year 2000. A national population projection of 280 million people in the year 2000 is divided among the market regions according to 1970 population proportions. The commodity restraints are then specified using per capita requirements for agricultural commodities in the year 2000 (Table 1.2).

The per capita demands for corn, sorghum, barley, oats, wheat, and sugar beets are based on average use from 1967 to 1969 for milling, brewing, and cereals. The per capita consumption level of cotton is projected to the year 2000 using the consumption levels over the past 30 years. The average sugar beet production per capita from 1967 to 1969 is used as a proxy for the demand for sugar. This procedure is used since a large proportion of the sugar consumed in the nation is imported from countries producing sugar cane, and to assume some increase in the proportion of total sugar from sugar beets is not warranted when compared to past trends in the sugar market.

Per capita consumption levels of beef, pork, and broilers are determined from price quantity equations. These equations were developed in a price-dependent form and for quantity determination were inverted to give:

$$Q_B = 43.7809 - 0.7697P_B + 0.1076P_{Br} - 0.0386Y \quad (1.11)$$
$$Q_P = 90.1111 - 0.2786P_B - 0.9612P_P + 0.0728P + 0.0032Y \quad (1.12)$$
$$Q_{Br} = 32.0623 + 0.1076P_B + 0.0728P_P - 0.4485P_{Br} + 0.0023Y \quad (1.13)$$

where Q_B is the beef consumed in pounds per capita in the year 2000 on a carcass weight basis; Q_P is the pork consumed in pounds per capita in the year 2000 on a carcass weight basis; Q_{Br} is the broilers consumed in pounds per capita in the year 2000 on a ready-to-cook basis; P_B is the expected price of beef in the year 2000; P_P is the expected price of pork in the year 2000; P_{Br} is the expected price of broilers in the year 2000; and Y is the projected per capita disposable income in the year 2000. Using the prices and disposable income projected in the year 2000, these equations are solved for the per capita consumption levels (Table 1.2).

Table 1.2. Projected per capita consumption levels for agricultural commodities in the year 2000

Commodity	Consumption	Commodity	Consumption
Corn	1.20 bushels	Fed Beef	108 lbs. carc. wt.
Sorghum	0.05 bushels	Nonfed Beef	51 lbs. carc. wt.
Wheat-Total	2.58 bushels	Dairy Products	4.04 cwt. milk eq.
Oats	0.22 bushels	Broilers[a]	50 lbs. ready to cook
Barley	0.58 bushels	Turkey[a]	9 lbs. ready to cook
Oilmeals[b]	0.09 cwt.	Lamb and mutton[a]	3 lbs. carc. wt.
Lint Cotton	12.00 pounds	Eggs[a]	207.5 eggs
Sugar beets	0.11 tons		

[a]Not used directly in the model's restraints, but used to determine the level of commodity demand and the resource use by class of livestock in the exogenous livestock sector.

[b]Oilmeal requirement reflects an adjustment for the high protein grain by-products provided from the milling of the per capita equivalent of the other grains.

The per capita consumption of turkeys, milk, eggs, and lamb and mutton are calculated from the following equations:

$$Q_T = e^{2.40871} P_T^{-0.43835} P_B^{0.19729} t^{0.21801} \tag{1.14}$$

$$Q_M = e^{6.6301 - 0.019t} \tag{1.15}$$

$$Q_E = e^{6.00183 - 0.1264t} \tag{1.16}$$

$$Q_L = e^{5.56087} P_L^{-1.9916} P_B^{0.57397} Y^{0.36813} t^{-.13775} \tag{1.17}$$

where Q_T is the turkey consumed in pounds per capita in 2000 on a ready-to-cook (RTC) basis; Q_M is the dairy products consumed in pounds per capita in the year 2000 on a whole milk equivalent basis; Q_E is the number of eggs consumed per capita in the year 2000; Q_L is the lamb and mutton consumed in pounds per capita in the year 2000 on a carcass weight basis; e is the base of the natural logarithm; P_T is the expected price of turkeys in the year 2000; P_B is the expected price of beef in the year 2000; t is the time in years after 1947; P_L is the expected price of lamb and mutton in the year 2000; Y is the projected per capita income in the year 2000.

The export/import levels used for all commodities are average annual volumes for 1969-71, (Table 1.3). A net import decreases the projection requirements of the commodity. Exports of corn grain, sorghum grain, barley, oats, wheat, and oilmeals are allocated to the market regions proportional to the average exports of each commodity from the major ports over the 1967-69 period (Nicol and Heady, 1975).

Table 1.3. Projected net exports of commodities for the year 2000

Commodity	Import	Export
	(thousand units)	
Corn		626,333 bu.
Sorghum		126,666 bu.
Barley		48,666 bu.
Oats		16,179 bu.
Wheat		958,719 bu.
Oilmeals[a]		276,407 cwt.
Cotton		3,306 bales
Beef	22,453 cwt	
Pork	3,349 cwt.	
Dairy Products[b]	4,661 cwt.	
Broilers		295,416 cwt.
Turkeys		44,162 cwt.
Eggs		68,699 doz.
Sheep and Lamb	1,647 cwt.	

[a]Oilmeals are expressed as soybean oilmeal equivalent of soybean oilmeal, cottonseed oilmeal, cottonseed and soybeans.

[b]Dairy products are expressed as cwt. of milk equivalents.

The Transportation Sector

The availability of and consumption requirements for commodities are defined by market region. This implies that there is no spatial differentiation among commodities produced or demanded in various producing areas within a market region. However, the cost of transporting commodities between market regions, is specified.

Transportation routes are defined between each pair of contiguous market regions. This transportation system is basically one of partial transshipment. However, some heavily used long-haul routes between noncontiguous regions also exist, and transportation routes are defined to represent the long-haul routes if the route reduced the mileage by 10 percent over the accumulated short haul routes. Over each route two activities are defined for each commodity, one activity for shipment in each direction.

A uniform rate is applied to each commodity over all routes. Ton/mile rates as functions of distance for various commodities are determined from data given in the 1966 Carload Waybill Statistics (Interstate Commerce Commission, 1968). Similar data on milk shipments are not available. However, over-the-road costs of fluid milk transportation have been estimated by Moede (1971).

In calculating the costs for the transportation activities, these rates are held constant. The carcass rate is used for both beef and pork. The cost for each activity is the distance of the route multiplied by the appropriate rate and converted into the units of the commodity restraint row.

A MATHEMATICAL EXPLANATION OF THE MODEL

A linear programming model forms a simultaneous equation system of restraint equations and an objective function to be optimized. This section outlines the objective function and the restraints that provide the interrelationships encompassed in this model.

The objective function, represented in equation (1.18), minimizes the cost of producing and then transporting the various crop and livestock commodities from producing areas to regions of consumption to meet domestic and export demands. All resources (except land and water) receive their market rate of return. The costs of water consumption and transfer also are included in the objective function. The activity costs cover all factors except land and water rents, which are reflected in shadow prices, and allow simulation reflecting the comparative advantage of each of the 223 producing areas and the 51 water supply regions, subject to environmental restraints and the level and location of consumer demands. The objective function is to minimize OF where:

$$\begin{aligned} OF = {} & \sum_i \left(\sum_k \left(\sum_m X_{ikm} UC_{ikm} + \sum_n Y_{ikn} UC_{ikn} \sum_m Z_{ikm}\right) + \sum_P L_{ip} UC_{ip}\right. \\ & + DPP_i UC_i + IPP_i UC_i + DWH_i UC_i + IWH_i UC_i + FLG_i UC_i \\ & + FP_i UC_i + \sum_w (WB_w UC_w + WD_w UC_w + WT_w UC_w) \\ & + \sum_t \sum_c T_{tc} UC_{tc} \end{aligned} \qquad (1.18)$$

The variables, parameters, and other terms are defined following equation (1.34).

Each of the 223 producing areas has land restraints by land group indicated in equations (1.19) to (1.25). Each region has a soil-loss restraint as in equation (1.26). A nitrogen balance equation as in equation (1.27) and a pasture restraint as in equation (1.28) are included in each region. Each water supply region has a water restraint as in equation (1.29), where the variables and parameters are defined subsequently.

Each of the 30 market regions has net demand equations for all of the relevant crop and livestock activities as illustrated by equation (1.30). Regional consumer demand quantities are determined exogenously from geographic and national projections of population, economic activity, per capita incomes, and international exports through the region for the year 2000. All technology and demand coefficients and the solution are for the year 2000. National demands are defined for cotton and sugar beets as indicated in equation (1.31). Poultry products, sheep, and other livestock are regulated at the consuming-region level. International trade is regulated at the region levels as indicated in equations (1.32) and (1.33).

Dryland cropland restraint, each region by land group:

$$\sum_m X_{ikm}a_{ikm} \leq LD_{ik} \tag{1.19}$$

Dryland wild hay restraint, each region:

$$DWH_ia_i \leq ADWH_i \tag{1.20}$$

Dryland permanent pasture restraint, each region:

$$DPP_ia_i \leq ADPP_i \tag{1.21}$$

Forestland grazed restraint, each region:

$$FLG_ia_i \leq AFLG_i \tag{1.22}$$

Irrigated cropland restraint, each region by land group:

$$\sum_n Y_{ikn}a_{ikn} + \sum_m Z_{ikm}a_{ikm} \leq LR_{ik} \tag{1.23}$$

Irrigated wild hay restraint, each region:

$$IWH_ia_i \leq AIWH_i \tag{1.24}$$

Irrigated permanent pasture restraint, each region:

$$IPP_ia_i \leq AIPP_i \tag{1.25}$$

Soil-loss restraint, each region, each land group, each activity:

$$SL_{ikm+n} \leq ASL_{ikm+n} \tag{1.26}$$

Nitrogen balance restraint, by region:

$$\begin{aligned} &FP_i + \sum_p b_{ip} L_{ip} + EL_{ic} b_{ic} - EC_i f_{ih} \\ &- \sum_k \left(\sum_m X_{ikm} f_{ikm} + \sum_n Y_{ikn} f_{ibn} + \sum_m Z_{ikm} f_{ikm} \right) \\ &- DPP_i f_i - IPP_i f_i - DWH_i f_i - IWH_i f_i - FLG_i f_i = 0 \end{aligned} \tag{1.27}$$

Pasture use restraint, per region:

$$\begin{aligned} &\sum_k \left(\sum_m X_{ikm} r_{ikm} + \sum_n Y_{ikn} r_{ikn} + \sum_m Z_{ikm} r_{ikm} \right) \\ &+ DPP_i r_i + IPP_i r_i + FLG_i r_i - \sum_p L_{ip} q_{ip} - EL_i q_i \geq 0 \end{aligned} \tag{1.28}$$

Water use restraint, by water region:

$$\begin{aligned} &WB_w \pm WT_w \pm WI_w - WO_w - WX_w - WE_w + WD_w - \sum_{i \epsilon w} IWH_i d_i \\ &- \sum_{i \epsilon w} IPP_i d_i - \sum_k \sum_{i \epsilon w} \left(\sum_m X_{ikm} d_{ikm} + \sum_n Y_{ikn} d_{ikn} \right. \\ &\left. + \sum_m Z_{ikm} d_{ikm} \right) - \sum_{i \epsilon w} \sum_p L_{ip} d_{ip} - \sum_{i \epsilon w} PN_i d_i \geq 0 \end{aligned} \tag{1.29}$$

Commodity balance restraints, each market region:

$$\begin{aligned} &\sum_k \sum_{i \epsilon j} \left(\sum_m X_{ikm} cy_{ikmc} + \sum_n Y_{ikn} cy_{iknc} + \sum_m Z_{ikm} cy_{ikmc} \right) \\ &\pm \sum_{i \epsilon j} \sum_p L_{ip} cy_{ipc} \pm \sum_{t \epsilon j} T_{tc} \pm E_{jc} - \sum_{i \epsilon j} PN_i cy_{ic} \\ &- EL_j cy_{jc} \geq 0 \end{aligned} \tag{1.30}$$

National commodity balance restraints for cotton, sugar beets, and spring wheat:

$$\begin{aligned} &\sum_i \sum_k \left(\sum_m X_{ikm} cy_{ikmg} + \sum_n Y_{ikn} cy_{ikng} + \sum_m Z_{ikm} cy_{ikmg} \right) \\ &- \sum_i PN_i cy_{ig} - EX_c \geq 0 \end{aligned} \tag{1.31}$$

National export restraints:

$$\sum_j E_{jc} \geq EX_c \tag{1.32}$$

National import restraints:

$$\sum_j E_{ic+e} \geq IM_{c+e} \tag{1.33}$$

Nonnegativity restraints:

$$X_{ikm}, Y_{ikn}, Z_{ikm}, L_{ip}, DWH_i, IWH_i, DPP_i, IPP_i, FLG_i, FP_i,$$
$$EL_l, WB_w, WT_w, WI_w, WD_w, WX_w, WE_w, PN_i, T_{tc},$$
$$E_{jc}, E_{icre} \geq 0 \tag{1.34}$$

The subscripts and variables for the equations are defined below:

Subscripts: c = 1, 2, ..., 15 for the endogenous commodities in the model; e = 1, 2, ..., 5 for the exogenous livestock alternatives considered; g = 1, 2, 3 for the commodities balanced at the national level; h = 1, 2, ..., 19 for the exogenous crop groups considered; i = 1, 2, ..., 223 for the producing areas of the model; j = 1, 2, ..., 30 for the market regions of the model; k = 1, 2, ..., 9 for the land groups in each producing area; m = 1, 2, ..., for the dryland crop management systems on a land group in a producing area; n = 1, 2, ..., for the irrigated crop management systems on a land group in a producing area; p = 1, 2, ..., for the livestock activities defined in the producing area; t = 1, 2, ..., 458 for the transportation routes in the model; and w = 1, 2, ..., 51 for the water supply regions in the model.

Variables: a = the amount of land used by the associated activity from the land base as indicated by the subscripts; AIPP = the number of acres of irrigated permanent pasture available in the subscripted producing area; ADWH = the number of acres of dryland wild (noncropland) hay available in the subscripted producing area; AFLG = the number of acres of forest land available for grazing in the subscripted producing area; ADPP = the number of acres of dryland permanent pasture available in the subscripted producing area; AIWH = the number of acres of irrigated wild (noncropland) hay available in the subscripted producing area; ASL = the per acre allowable soil loss subscripted for land group, producing area, and activity; b = the units of nitrogen-equivalent fertilizer produced by livestock, subscripted for producing area and activity; cy = the interaction coefficient of the relevant commodity as regulated by the associated activity and specified by the subscripts; d = the per unit of activity water use coefficient as regulated by the associated activity and specified by the subscripts; DPP = the level of use of dryland permanent pasture in the subscripted producing area; DWH = the level of use of dryland wild (noncropland) hay in the subscripted producing area; E = the level of net export for the associated commodity in the associated region as specified by the subscripts; EC = the level of exogenous crop production by subscripted region; EL = the level of exogenous livestock production consistent with the subscripted region; EX = the level of national net export for the subscripted commodity as determined exogenously to the model; f = the units of nitrogen-equivalent fertilizer required by the associated activity and specified by the subscripts; FLG = the level of forestland grazed in the subscripted producing area; FP = the number of pounds of nitrogen-equivalent fertilizer purchased in the subscripted producing area; IM = the level of national net imports for the subscripted commodities as determined exogenously to the model; IPP = the level of use of the irrigated permanent pasture in the subscripted region; IWH = the level of the irrigated wild (noncropland) hay in the subscripted region; L = the level of the livestock activity with the type and region dependent on the subscript; LD = the number of acres of dryland cropland available for use as specified by the region and land group subscripts; LR = the number of acres of irrigated cropland available for use as specified by the region and land group subscripts; PN = the level of population projected to be in the subscripted region; q = the units of pasture, in hay equivalents, consumed by the associated livestock activity and specified by the subscripts; r = the units of aftermath or regular pasture, in hay equivalents, produced by the associated cropping or pasture activity and identified by the subscripts; SL = the level of soil loss associated with any activity over the range m + n in the region and land group designated by the subscripts; T = the level of transportation of a unit of the commodity either into or out of the market region designated by the subscripts; WB = the level of water purchase for use in the water balance of the water supply region designated by the subscript; WD = the level of desalting of ocean water in the water supply region designated by the subscript; WE = the level of water to be exported from the water supply region subscripted; WI = the level of movement of water in or out of the water supply region through the interbasin transfer network; WO = the level of water requirement for onsite uses such as mining, navigation, and estuary maintenance in the water supply region subscripted; WX = the level of water use for the exogenous agricultural crops and livestock in the water supply region subscripted; X = the level of employment of the dryland crop management system in the region and on the land group as designated by the subscripts; Y = the level of employment in the irrigated crop management system in the region and on the land group as designated by the subscripts; and Z = the level of employment of the dryland crop management system on the land group in the region as designated by the subscripts when the land has been designated as available for irrigated cropping patterns.

Included in a solution of the model is an analysis of the restraints. The restraint analysis of the solution indicates the level of use and the implied value of the restraint.

The level of use of a restraint is determined as the sum of all activity levels multiplied by their interaction coefficient for the given restraint or:

$$RL_j = \sum_i X_i a_{ij} \tag{1.35}$$

where RL_j = the level of use of resource j; X_i = the level of activity i; and a_{ij} = the interaction coefficient between activity i and resource j. The value of RL_j will be confined by the lower and upper limits specified for the resource.

The implied value of the resource is often referred to as the shadow price and represents the marginal value product of the resource. The marginal value product represents the amount by which the total cost (objective function value) of the program will be reduced if the limit on the supply of the resource is reduced by one unit. In other words, it represents the marginal value product of the last unit of the resource used. On the restraints that simulate markets, this represents the marginal cost of the last unit of the commodity produced and reflects the market price of all production and transportation costs considered in the model.

Within the model, as developed, shadow prices are determined at the producing-area level for nitrogen fertilizer, pasture (in hay-equivalent value), and land by the 9 land groups. Shadow prices are determined at the water-supply-region level for water and at the market-region level for all the commodities except sugar beets, cotton, and spring wheat, which have national markets and, thus, national shadow prices.

POLICY ALTERNATIVES AND ANALYSIS

To evaluate the impacts in the agricultural sector from imposed limits on soil erosion losses, alternative sets of assumptions concerning allowable soil-loss and export levels are entered into the model. A summary of the alternatives evaluated is given in Table 1.4. Alternatives B, C, and D, compared to Alternative A, allow less and less soil loss while holding export levels constant. Alternatives E and F, compared to Alternative C, vary export levels in the model while holding the allowable soil loss to 5 tons per acre (Table 1.5). These alternatives represent a finite number of the complete set of possible alternatives. It is hoped that the analysis of these alternatives identifies trends and tradeoffs that need to be examined when controls on erosion are considered for legislation.

Table 1.4. Soil-loss and export assumptions for the alternatives

Alternative	Per Acre Allowable Soil Loss Level	Export level[a]
A	Unrestricted	1969-71 average
B	10 tons	1969-71 average
C	5 tons	1969-71 average
D	3 tons	1969-71 average
E	5 tons	Double 1969-71 average
F	5 tons	Triple 1969-71 average

[a]Quantities by commodity are shown in Table 1.5.

Table 1.5. Export levels of feed grains, wheat and oilmeals in 2000

Commodity (units)	1973[a]	Alternative A,B,C, and D[b]	E	F
	(thousand units)			
Corn (bu.)	1,250,000	626,333	1,252,666	1,878,999
Sorghum (bu.)	190,800	126,666	253,332	379,998
Barley (bu.)	666,300	48,666	97,332	145,998
Oats (bu.)	23,500	15,666	31,332	46,998
Wheat (bu.)	1,847,700	658,719	1,317,438	1,976,157
Oilmeals (cwt.)	276,407	552,814	829,218	317,400

[a]Exports during the 1972-73 crop year.

[b]Average export levels for the years 1969-71.

Agriculture and Environmental Improvement

Varying the per acre soil-loss restrictions in the model indicates that agriculture can contribute to environmental improvement through reduced soil erosion without serious implications for domestic food prices when exports are at the 1969-71 levels. It is expected that this reduced erosion will lower agriculture's contribution to the level of sedimentation in the nation's waterways. There is, however, a tradeoff in environmental quality implied by the results: as erosion is reduced, the levels of pesticide and fertilizer application increase, as farmers shift to reduced tillage methods.

Raising commodity export levels in the model results in increased soil erosion from 2.8 to 2.9 tons per acre in the model's solution to Alternative C. Soil loss increases because a higher proportion of the more erosive lands in the land base are brought into production to meet the expanded demands. Total soil erosion increases 14.0 percent and 43.0 percent, respectively, as exports double and triple, even though the model is constrained to a limit of 5 tons per acre for each alternative (Table 1.6).

Table 1.6. Summary of erosion losses and management practices with the alternative soil loss restrictions and export levels in the year 2000

Item (unit)	Alternatives					
	A	B	C	D	E	F
Erosion Per Acre (tons)	9.9	4.3	2.8	1.9	2.9	2.9
Total Erosion (billion tons)	2.7	1.1	0.7	0.5	0.8	1.0
Total Acres Cultivated (million)	269	262	259	258	295	337
Percent of Acres by Tillage Method:						
Conventional	92.1	83.1	77.7	71.7	75.2	72.9
Reduced	7.9	16.9	22.3	28.3	24.8	27.1
Percent of Acres by Conservation Practice:						
Straight row	94.2	74.0	59.5	48.8	54.8	49.4
Contouring	4.6	17.3	21.6	25.2	23.3	23.4
Strip crop/terracing	1.2	8.7	18.9	26.0	21.9	27.2

Alternative Erosion Control Methods

Erosion apparently can be controlled by currently available technologies with little impact on agriculture's potential to meet both domestic agricultural demands in the year 2000 and modest export increases. Results indicate that a significant level of erosion control in crop production might be attained through a rather large shift to reduced tillage methods and contouring and strip cropping.

Table 1.6 presents a summary of the soil-loss results and farming practices employed for the alternatives analyzed. Erosion losses per acre with no restriction average 9.9 tons per acre and decline steadily with increasingly stringent controls to 1.9 tons per acre with a 3-ton restriction. Total soil loss also drops markedly from the 2.677 billion tons under Alternative A. The methods of controlling erosion to bring about these large reductions in soil loss include shifts from conventional tillage to reduced tillage and from straight-row cultivation to other conservation practices. (See Table 1.6.) Under Alternative A, 92.1 percent of the lands are farmed using conventional tillage methods. This drops to 71.7 percent for the 3-ton restriction under Alternative D. With respect to conservation practices, 94.2 percent of the land is cultivated using straight-row farming practices when no soil-loss restriction is imposed. With Alternative D, the percent of total cropped acres under straight-row farming practices declines to 48.8 percent. Concurrently, the use of contouring to reduce soil erosion increases from 4.6 percent under Alternative A to 25.2 percent of the total cropland acreage under Alternative D.

In addition to the changes in the farming practices and conservation methods, the changes in allowable soil losses alters both the mix of crops grown and the land group upon which specific crops are grown. As shown in Table 1.6, the total acres of row crops declines while the acreage of hay increases. This change in crops is possible because of substitutions between feedstuffs in the livestock rations, e.g., less silage and more hay. The decrease in silages is due to relative price changes in feedstuffs. Silages become a relatively higher-priced feedstuff because their very erosive characteristics require high-cost practices to protect the land from soil loss. Besides the adjustment in total acres of some crops because of substitutions in livestock rations, there is a significant shift in the use of the land, by land capability class, in the production of these crops. The percentage of total acres of row crops grown on the lands least subject to erosion increases. (See classes I and II in Table 1.7.) The opposite occurs on the Class III e and IV e lands, which are very susceptible to erosion. These shifts in the use of land contribute significantly to reduced erosion.

Table 1.7. Percentage breakdown by crop categories of total dryland acreage for soil capability classes

		Percent of U.S. Total by Soil Capability Class[a]			
Crop Type	U.S. Total	I-II	III_e-IV_e	Other III-IV	V-III
Alternative A					
Row crops	138,980	73	16	11	—
Close grown crops	70,205	55	31	13	—
All hay	66,333	44	21	9	—
Alternative B					
Row crops	132,636	77	12	11	—
Close grown crops	68,911	56	30	13	—
All hay	67,289	42	24	9	—
Alternative C					
Row crops	128,505	78	11	11	—
Close grown crops	66,732	56	30	16	—
All hay	72,508	43	21	11	—
Alternative D					
Row crops	127,925	80	9	11	—
Close grown crops	66,828	55	33	12	—
All hay	72,858	46	26	3	—
Alternative E					
Row crops	152,212	73	13	14	—
Close grown crops	76,537	52	35	13	—
All hay	75,800	42	23	11	—

[a]Soil Capability classes as defined by Soil Conservation Service for the Conservation Needs Inventory (U.S. Department of Agriculture, 1971).

Regional Cropping Patterns

Imposing soil-loss restrictions also alters the mix of crops grown in each region consistent with the susceptibility of that region's cropland to erosion. To continue growing erosive crops in areas very susceptible to erosion requires more costly management practices relative to areas not subject to high erosion losses. Within the framework of interregional competition, these relative changes in production costs shift the more erosive crops to areas with fewer erosion problems. Thus the imposition of soil-loss restrictions reduces the acres of row crops and close-grown crops in the South Atlantic, South Central, and North Central regions (Table 1.8). To compensate, the total acres of hay and pasture increase in the

Table 1.8. Summary of agricultural land use by region under alternative soil loss restriction levels in 2000

	Alternative			
Region/Land Use	A	B	C	D
North Atlantic				
Row crops	3,713	3,661	4,318	3,359
Close crops	1,698	1,225	2,491	2,089
Hayland	1,616	1,608	2,042	2,247
Pasture	6,808	6,888	7,098	8,315
South Atlantic				
Row crops	10,691	9,647	10,065	8,549
Close crops	3,201	2,729	870	736
Hayland	2,007	2,006	2,686	2,913
Pasture	23,592	23,632	25,222	25,615
North Central				
Row crops	90,926	87,843	84,714	88,372
Close crops	19,440	19,117	18,756	17,652
Hayland	17,190	18,054	20,421	23,523
Pasture	29,775	30,304	33,820	35,810
South Central				
Row crops	28,666	26,401	24,305	19,883
Close crops	20,716	20,569	19,720	19,885
Hayland	27,447	27,648	27,722	23,304
Pasture	110,694	111,368	108,031	105,322
Great Plains				
Row crops	7,512	7,017	5,593	6,904
Close crops	16,192	16,549	14,410	18,292
Hayland	21,051	21,385	22,443	24,775
Pasture	41,860	41,949	41,999	45,579
North West				
Row crops	1,740	1,734	1,981	1,811
Close crops	7,763	7,762	7,751	7,906
Hayland	5,679	5,631	5,598	5,986
Pasture	26,896	27,033	27,833	28,872
South West				
Row crops	4,978	5,112	5,059	5,562
Close crops	6,498	6,358	8,380	6,253
Hayland	3,747	3,737	3,689	2,826
Pasture	66,357	66,361	67,404	66,925

North Central and South Atlantic regions. The national interregional adjustment to these changes is that in the more arid regions, including the Great Plains, Northwest, and Southwest, there are increase in close-grown crops. The Northwest and Southwest regions also show increases in row-crop acreages.

Increased exports necessarily require more land to be cultivated and/or a more intensive use of the lands already farmed. The extra feed grains, wheat, and oilmeals for export in Alternatives E and F directly increase the acreage of the row crops and close-grown crops in the model (Table 1.9). The increase in the use of water and nitrogen fertilizer at the national level is nearly in proportion to the increased acreage needed for the higher exports. There are regional variations from proportionate increases due to limits on available water and irrigable land and the crop-yield response to fertilizers used in the various regions and on the several land groups.

Table 1.9. Acreage of cultivated land by region and alternative export levels

Region	Alternatives C	E	F
	(thousand acres)		
National[a]	258,882	295,049	337,299
North Atlantic	6,680	8,187	19,742
South Atlantic	10,225	14,401	18,088
North Central	114,340	127,725	138,711
South Central	66,142	70,409	78,705
Great Plains	35,367	46,409	59,437
North West	11,603	12,357	14,494
South West	14,487	15,561	17,055

[a]Total does not sum due to rounding.

Commodity Price Levels

If the unrestricted soil-loss alternative, Alternative A, is considered a base, then comparing the other solutions with Alternative A provides an indication of the impact the restrictions would have on food prices–a measure consumers must weigh when evaluating the benefits and burdens of each alternative. Given agricultural capacity as defined in the model and the 1969-72 export levels, a reduction in soil loss can be attained with only minor increases in prices (Table 1.10). The 5-ton per acre soil-loss restriction results in an average increase in supply prices of 5 percent when exports are held constant at the 1969-71 levels. If exports of the feed grains, wheat, and oilmeals are increased while continuing to impose the 5-ton soil-loss restriction, however, price increases are greater (Table 1.11). Output cannot be significantly increased without substantial increases in prices, as much of the additional land brought into crop production is less productive and in some cases requires expensive soil-conserving production practices to meet the soil-loss limitation.

Table 1.10. Indices of supply prices for selected agricultural commodities under alternative soil loss restrictions

	Alternatives			
Commodity	A	B	C	D
Corn	100	100	107	106
Wheat	100	99	103	103
Soybeans	100	101	115	121
Cotton	100	100	112	125
Hay	100	99	101	106
Beef	100	100	104	105
Pork	100	100	105	104
Milk	100	100	100	102

Table 1.11. Indices of supply prices for selected agricultural commodities with a 5 ton soil loss restriction and alternative levels of exports

	Alternatives		
Commodity	C	E	F
Corn	100	105	126
Wheat	100	109	158
Soybeans	100	139	219
Cotton	100	101	107
Hay	100	109	132
Beef	100	106	119
Pork	100	106	122
Milk	100	103	111

Single Region Analysis

Analysis of income variations of a single market region gives insight into possible impacts and changes under the soil-loss restrictions. Market region 21, centered around Lincoln, Nebraska, is used as an illustration. This market region does not show large shifts in commodities produced or in technologies employed under either Alternative C or D compared to Alternative A. However, a reduction of 11 percent in income under Alternative C and an increase of 9 percent under Alternative D illustrate the importance of the level of the environmental restraint on the region.

Table 1.12 gives details of changes in commodity production and income for this region. Proportions of total income derived from each of the commodities varies significantly across alternatives. For example, production of silage is highly erosive. Therefore, silages are not used as much under Alternatives C and D as with Alternative A. Income from silage production as a proportion of total income decreases to only 4 percent of total regional income in Alternative D as production is reduced 57 percent from Alternative A. This follows the national trend in silage production with these alternatives. Since silage does not have a final consumer demand, it is replaced by grains in livestock rations, resulting in increased corn production. The largest component of the 11 percent decline in

Table 1.12. Income data by commodity for Market Region 21 for selected alternatives

	Alternative A		Alternative C			Alternative D		
Commodity	Gross farm income[a]	Percent of total income	Gross farm income[a]	Percent of total income	Percent change from Alt. A	Gross farm income[a]	Percent of total income	Fercent change from Alt. A
Corn	118,816	7	192,287	12	62	102,211	5	14
Sorghum	6,405	0	5,289	0	-17	55,596	3	768
Barley	1,454	0	1,100	0	-24	2,245	0	54
Oats	7,493	0	8,694	1	16	11,699	1	56
Wheat	39,351	2	20,235	1	-49	71,558	4	82
Oilmeals	53,805	3	45,235	3	-16	99,891	5	86
Legume Hay	173,299	10	160,854	10	-7	265,407	14	53
Nonlegume Hays	119,887	7	142,163	9	19	131,535	7	10
Silage	179,915	10	75,592	5	-58	76,853	4	-57
Feeders	599,606	33	461,637	29	-23	557,330	28	-7
Fed Beef	325,299	18	354,719	22	9	426,214	22	31
Nonfed Beef	158,618	9	121,528	8	-23	144,963	7	-9
Total Income[b]	1,795,580		1,594,241		-11	1,964,703		9

[a]In thousands of dollars.

[b]Greater than sum of columns due to commodities not listed.

market region 21's income for Alternative C is a shift away from feeder cattle, which are produced in other regions at lower cost.

For alternative D, income from grains other than corn increases dramatically. This increase accompanies a substantial increase in fed beef and a decrease (from Alternative A) in income from feeders. As previously noted, the erosion limitation has a pronounced effect on the agricultural production of the Southeast, particularly the South Central and Mississippi Delta states. Crops and technologies available in these regions are highly restrained. Thus the row-crop production of feed grains and soybeans must shift to other regions. The Midwest and Great Plains areas benefit from this shift. Market region 21 has a strong economic advantage in feed-grain and wheat production and thus produces more of these commodities to meet national demands. Legume hay production increases in market region 21 for Alternative D, after a precipitous decrease in this commodity for Alternative C. This increase corresponds to greater production of feeders under Alternative D.

Shifts in Returns to Resources and Income Distribution

Imposing soil-loss restrictions on the model differentially alters returns to land resources. In general, the more productive lands have increased returns (shadow prices imputed by the model) while the less productive lands have reduced returns (Table 1.13). Based on national averages from the model solutions, the return to Class I and Class II land increases from $20.60 per acre under Alternative A to $25.30 under Alternative D. The average return to lands in classes V-VIII, however, drops from $10.87 per acre to $7.50 per acre, as the allowable soil-loss level is reduced to 3 tons per acre. The increased exports under Alternative E produce an even greater differential effect by land group. The price imputed by the model under Alternatives E and F to Class I and II land is significantly higher than in Alternative C, while the price for Class V-VIII land

Table 1.13. Shadow prices of land groups by alternative for the year 2000

Alternative	Land capability classes[a]							
	I-II		$III-IV_e$		$III-IV_e$		V-VIII	
	$/acre	index	$/acre	index	$/acre	index	$/acre	index
A	20.60	100.0	10.67	100.0	12.02	100.0	10.87	100.0
B	20.01	97.6	9.95	93.2	10.43	86.8	5.68	52.2
C	22.47	109.1	9.06	84.9	9.72	80.9	6.72	61.8
D	25.30	122.8	11.93	111.8	11.24	93.5	7.05	64.8
E	32.25	156.6	14.00	131.0	15.22	126.6	2.06	18.9

[a]Land capability classes as defined by the Soil Conservation Service for the Conservation Needs Inventory (U.S. Department of Agriculture, 1971).

is significantly lower. A relative shift in asset values would occur among the owners of agricultural lands with the imposition of soil conservation programs.

Another equity aspect of soil-loss control programs is that as increasingly stringent soil loss allowances are imposed, more and more land in regions highly subject to erosion is idled. Nationally, 11 million acres are taken out of production in Alternative D, compared to Alternative A (Table 1.6). On a regional basis, the results are even more dramatic. The idling of croplands is especially severe in the South Central region, where over 14 million acres of land are taken out of production when the 3-ton limit is imposed on the model. The South Atlantic region has over 4 million acres of land idled under the 3-ton restriction. Regions not as subject to erosion gain under these restrictions. The Great Plains region and North Central region have additional acres under cultivation in Alternative D—5 million acres and 2 million acres, respectively, compared to Alternative A.

The data in Table 1.14 indicate the total value of resources utilized in agricultural production under each alternative in each of the 7 reporting regions. Land and water resources are valued according to the value of the last unit utilized (the shadow prices in the solution). The labor, machinery, and pesticide costs in the regional totals are the component costs in the cropping-system budgets. The cost of fertilizers and the associated miscellaneous items of production are also included. The South Atlantic and South Central regions use

Table 1.14. The value of resources employed by the agricultural sector of each region and percent of national total[a]

	Alternative									
	A		B		C		D		E	
	(percent)									
North Atlantic	456	3.2	434	3.1	623	4.3	490	3.3	782	4.2
South Atlantic	923	6.5	863	6.2	854	6.0	740	4.9	1,318	7.0
North Central	7,891	55.7	7,840	56.0	7,973	55.7	8,708	58.2	10,212	54.5
South Central	2,919	20.6	2,881	20.6	2,951	20.6	2,651	17.7	3,704	19.7
Great Plains	1,061	7.5	1,054	7.5	916	6.4	1,130	7.6	1,556	8.3
North West	410	3.0	409	2.9	427	3.0	453	3.0	554	2.9
South West	513	3.6	519	3.7	570	4.0	779	5.2	625	3.3
United States	14,163	100.0	14,000	100.0	14,314	100.0	14,951	100.0	18,750	100.0

[a]Values are in 1972 dollars.

fewer resources as the soil-loss restriction reduces cropped acreages. The North Central and Great Plains regions utilize more resources as acreages expand and the more intensive conservation practices require more inputs, especially pesticides. The value of resources used in the 2 regions in the West increase, especially as the soil-loss restriction reaches the 3-ton level. Increasing export levels from Alternative C to Alternative E results in a greater use of agricultural inputs but little change in each region's proportion of the national total.

SUMMARY

Soil erosion can be reduced significantly with only small increases in commodity prices with low exports. Total soil loss can be reduced through (1) increased use of contouring and strip cropping and reduced tillage; (2) growing fewer acres of the more erosive crops, especially the silages; and (3) incorporating the more erosive crops into less intensive row-crop rotations. To substantially increase exports while continuing to impose very stringent soil-erosion restrictions, however, will raise commodity prices sharply.

Soil erosion control restrictions give pronounced shifts in production patterns and resource use in the South Central, South Atlantic, Northwest, and North Central regions. Regions in the West, where runoff is lower, gain in production of agricultural products. In addition to these interregional consequences, landowners with the more productive lands less subject to erosion receive higher returns, while the owners of the less productive land subject to erosion receive lower returns.

REFERENCES

Dean, W. G., G. A. King, H. O. Carter, and C. R. Shumway. 1970. Projections of California Agriculture to 1980 and 2000. California Agricultural Experiment Station Bulletin 847. University of California, Berkeley.

Economic Research Service. 1971a. *Selected U.S. Crop Budgets: Yields, Inputs, and Variables Costs*. Vol 3: *Great Plains Region*. ERS459. USDA, Washington, D.C.

___. 1971b. *Selected U.S. Crop Budgets: Yields, Inputs, and Variables Costs.* Vol. 5: *South Central Region* ERS461. USDA, Washington, D.C.

Eyvindson, R. H. 1965. A Model of Interregional Competition in Agriculture Incorporating Consuming Regions, Producing Areas, Farm Size Groups, and Land Classes, vols. 1-5. Unpublished Ph.D. dissertation, Iowa State University, Ames.

Heady, E. O., and L. V. Mayer. 1967. Food Needs and U.S. Agriculture in 1980. Technical Papers, vol. 1. National Advisory Commission on Food and Fiber. Washington, D.C.

Howe, C. W., and K. W. Easter. 1971. *Interbasin Transfers of Water-Economic Issues and Impacts*. Johns Hopkins Press, Baltimore.

Ibach, D. B., and J. R. Adams. 1967. Fertilizer Use in the United States by Crops and Areas, 1964 Estimates. USDA Statistical Bulletin 408. Economic Research Service and Statistical Reporting Service, Washington, D.C.

Interstate Commerce Commission. 1968. Carload Waybill Statistics, 1966 Mileage Block Distribution: Traffic and Revenue by Selected Commodity Classes, Territorial Movement, and Type of Rate. U.S. Government Printing Office, Washington, D.C.

Jennings, R. D. 1955. Relative Use of Feeds for Livestock Including Pasture, by States. USDA Statistical Bulletin 153. Agricultural Research Service, Washington, D.C.

Lof, G. O. G., and C. H. Hardison. 1966. Storage Requirements for Water in the United States. *Water Resources Research* 2(3): 323-354.

Moede, H. H. 1971. Over-the-Road Costs of Hauling Bulk Milk. Marketing Research Report 919. USDA Research Service, Washington, D.C.

National Academy of Sciences. 1968. Nutrient Requirements of Swine, 6th rev. ed. NAS, Washington, D.C.

___. 1970. Nutrient Requirements of Beef Cattle, 4th rev. ed. NAS, Washington, D.C.

___. 1971. Nutrient Requirements of Dairy Cattle, 4th rev. ed. NAS, Washington, D.C.

Nicol, K. J., and E. O. Heady. 1975. A Model for Regional Agricultural Analysis of Land and Water Use, Agricultural Structure, and the Environment: A Documentation. Miscellaneous Report. Center for Agricultural and Rural Development, Iowa State University, Ames.

Shipley, J. L., and J. E. Osborn. 1973. Costs, Inputs, and Returns in Arid and Semiarid Areas. Conservation Tillage. Proceedings of a national conference. Soil Conservation Society of America, Ankeny, Iowa.

Shrader, W. D., and R. D. Voss. 1972. Soybeans may supply 50 pounds of nitrogen for next year's corn. *Wallaces' Farmer* 97(Sept. 23):42-43.

Soil Conservation Service, Engineering Division. 1972. Procedure for Computing Sheet and Rill Erosion on Project Areas. Soil Conservation Service Technical Report 51. USDA, Washington, D.C.

Stoecker, Arthur. 1974. A Quadratic Programming Model of U.S. Agriculture in 1980: Theory and Application. Unpub. Ph.D. diss., Iowa State Univ., Ames.

U.S. Department of Commerce, Bureau of Census. 1968a. *U.S. Census of Agriculture, 1964*. Vol. 1: *Statistics for the States and Counties*. U.S. Government Printing Office, Washington, D.C.

___. 1968b. *U.S. Census of Agriculture, 1964.* Vol. 2: *General Report*. U.S. Government Printing Office, Washington, D.C.

U.S. Department of the Interior, Bureau of Land Management. 1970. Use of Water on Federal Irrigation Projects. Interim Report. Denver, Colorado.

U.S. Water Resources Council. 1968. The Nation's Water Resources. U.S. Government Printing Office, Washington, D.C.

___. 1970. Water Resources Regions and Subregions for the National Assessment of Water and Related Land Resources. U.S. Government Printing Office, Washington, D.C.

Waugh, F. V. 1964. Demand and Price Analysis. Economic and Statistical Analysis Division. USDA Technical Bulletin 1316. Washington, D.C.

Wischmeir, W. H., and D. D. Smith. 1965. Predicting Rainfall-Erosion Losses from Cropland East of the Rocky Mountains. USDA Agriculture Handbook 282. Washington, D.C.

CHAPTER 2. INTERREGIONAL ANALYSIS OF SOIL CONSERVANCY AND ENVIRONMENTAL REGULATIONS IN IOWA WITHIN A NATIONAL FRAMEWORK

by Vishnuprasad Nagadevara and Earl O. Heady

INTENSITY of public concern over soil loss and land use increased in Iowa in 1971 when the state legislature created conservancy districts within which maximum soil-loss limits were established on agricultural lands at 1 to 5 tons per acre per year, depending on the soil type. (This law has now been implemented.) Such isolated legislation or controls on allowable soil loss or use of other agricultural inputs in any one state could cause undesirable effects on the well-being of farmers in that state. These isolated measures to improve the environmental quality of a particular state can make it less profitable to use the land and other inputs in that state compared to the rest of the country. This profit reduction, in turn, might reduce land values and decrease the net worth of the local farming sector. In addition, if the state in which the controls are imposed is large enough and production and prices are altered sufficiently, other states may gain from increased incomes and higher land values.

This study was made under a grant from the National Science Foundation's Research Applied to National Needs (RANN) program (Nagadevara et al., 1975). An interregional linear programming model was formulated to simulate the effects of environmental controls enacted and practiced in Iowa but not elsewhere in the nation. This study was conducted to answer questions such as: Will environmental improvement measures enacted in a single state such as Iowa cause its producers to gain or sacrifice income as their land use and farm practices are restricted? What would be the impact on consumers? How are these impacts altered as export levels change?

The linear programming model employed in the analysis is a modification of the linear programming model described in Chapter 1. The discussion about the model which follows is brief, and the reader is referred to Chapter 1 for details.

THE MODEL

This section provides a brief overview of the model and summarizes the modifications of the linear programming model described in the previous chapter to allow the investigation of the impact of the environmental controls on Iowa agriculture. The model has four major parts: (1) the land and water resources available to agriculture by region, (2) crop and livestock production activities for the transformation of these resources into agricultural commodities, (3) the interregional commodity transportation network, and (4) the domestic and foreign demands for agricultural products. The model is solved with the objective of meeting the demands for agricultural products in a manner that will minimize the total cost of producing and transporting the nation's agricultural products.

Regions of the Model

Four separate sets of regions are defined in this model. The first set is the producing areas within which the land resources and the associated production activities are defined. These differ from the 223 producing areas in Chapter 1. The second set, referred to as water supply regions, specify water availability and water transfer possibilities. These are the same as the regions in Figure 1.2. The third set of regions are market regions within which the markets are defined; they differ somewhat from those in Figure 1.3. The fourth set represents the regions into which the results from the producing areas and market regions are aggregated for reporting. The only reporting regions are Iowa and the United States.

The producing areas defined for U.S. agriculture in this model are shown in Figure 2.1. Ninety of these producing areas are based on the county approximations of the 206 subareas defined by the Water Resources Council (1970) and parallel those in Figure 1.1 for producing regions. Some of the producing areas in figure 2.1 are aggregations of those in Figure 1.1. Twelve producing areas are used to define the agricultural lands and the cropping systems in Iowa, as shown in Figure 2.2. These Iowa producing areas are consistent with the Iowa soil conservancy districts

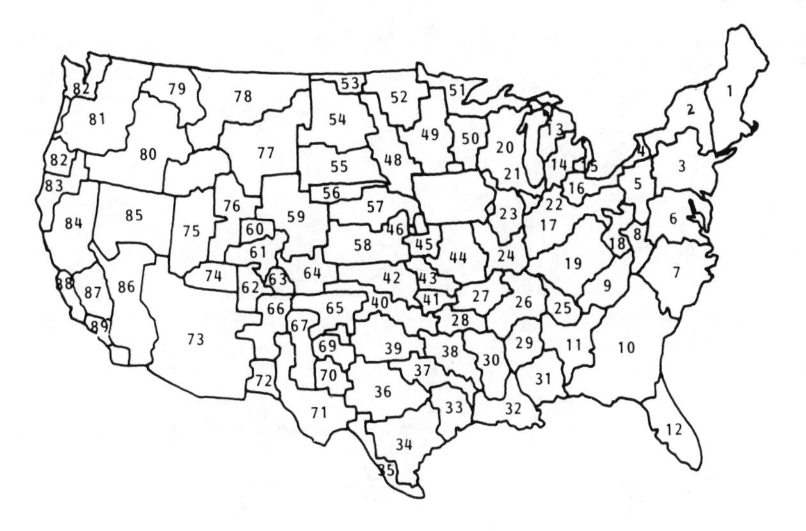

Figure 2.1. The producing areas for the rest of the United States.

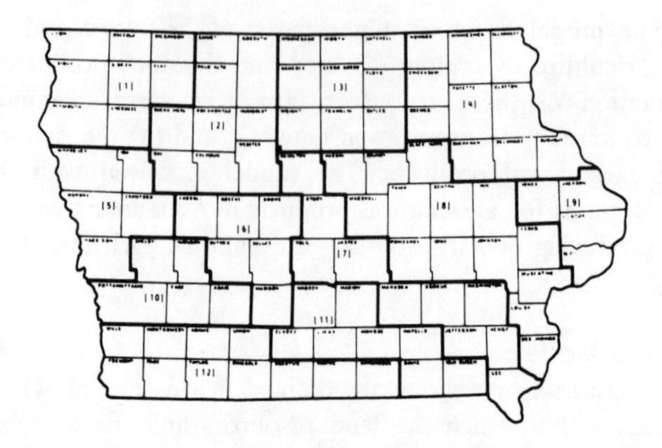

Figure 2.2. Iowa's 12 conservancy-producing areas.

for which the analysis is made (Cooperative Extension Service, 1972). These 12 producing areas are aggregations of contiguous counties within soil conservancy districts as defined by Iowa law.

Each water supply region is defined to approximate a physical region within which a water supply can be said to exist. These water supply regions are aggregations of contiguous producing areas. Subdivisions of the 18 major river basins of the Water Resources Council are the basis of these regions. (The water regions for this study are the same as those shown in Figure 2.2.) The activities in the model creating the demand for the supply of water, along with buying and transportation activities, are defined within these regions as explained in Chapter 1.

Twenty-nine market regions are defined in the model (Figure 2.3). They differ

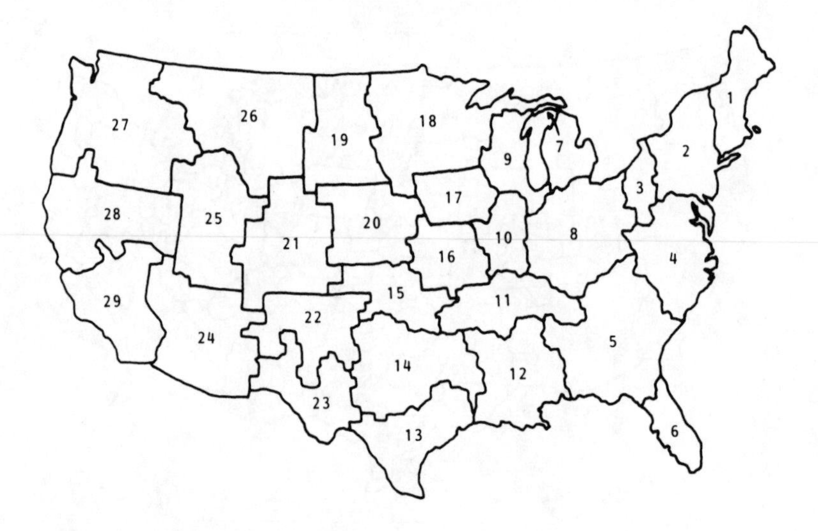

Figure 2.3. The 29 market regions.

from the 30 in Figure 1.3 due to the particular specifications of the Iowa model. These regions are again delineated around major metropolitan areas of the United States. Each market region is an aggregation of contiguous producing areas. Metropolitan centers are identified in each market region to serve as the links of the model's transportation sector and in some cases ports for exports.

The Land Base

The model's land base was built from the Conservation Needs Inventory (CNI), which reports acres of land by use and by agricultural capacity class (USDA, 1970) as explained in Chapter 1. Also explained in Chapter 1, the CNI major capability classes were used to formulate 9 land groups, as shown in Table 1.1, for each producing region. The county acreages were summed by dryland and irrigated uses to the 102 producing areas using the twenty-nine capability class-subclass categories (Table 1.1).

The Crop Production Sector

The same endogenous and exogenous crops were used for this model emphasizing Iowa as for the model in Chapter 1. Similarly, three tillage methods and four conservation practices were included for each land group in each producing area. Each rotation is defined with one of four conservation practices: straight-row cropping, contouring, strip cropping, or terracing. The definition of a crop management system is completed by specifying one of three tillage practices: conventional tillage with residue removed, conventional tillage with residue left, or reduced tillage. These crop management systems, which become the activities in the model's crop production sector, are adjusted to account for differences in production cost, fertilizer requirements, crop yields, water needs, and susceptibility to soil erosion in each producing area. The procedure used to specify coefficients for crop rotations allows for interrelationships among crops, e.g., nitrogen carryover following legume crops. Using the land and water resources defined in the model and fertilizer, each activity produces commodities needed for livestock and consumer demands.

The Livestock Production Sector

The livestock sector is the same as that defined in Chapter 1. Coefficients for feed requirements and manure production are specified for all categories of livestock, but cost data are needed only for the endogenous livestock production activities.

Livestock rations are formulated to allow substitution between grains, between roughages and grains, and between roughages. Hence, the model selects least-cost rations for each class of endogenous livestock in each region. The nitrogen in the manure produced by the livestock sector is transferred to the crop production sector where it is utilized as a fertilizer.

The Demand Sector

The demand sector of the model specifies the projected levels of domestic requirements for food and fiber, net exports, exogenous livestock feed requirements, and the industrial and nonfood uses. These demands are defined in each market region. The projected domestic demand levels for food and fiber and the projected net exports for this analysis are shown in Tables 2.1 and 2.2, respectively. An explanation of the alternative levels in the demand sector is provided in the next section. Some of the lower levels of demand are the same as those shown in Tables 1.2 and 1.3 of the previous chapter. However, Tables 2.2 and 2.3 also employ some higher export levels as a means for exploring their impacts on soil erosion, commodity prices, resource returns, and related values.

Time Horizon

Evaluation of policy impact alternatives within the limitations of the model requires that a sufficient time horizon be specified to allow for the implied adjustments to materialize. In this study, 1985 was selected as the year of analysis. The alternative formulations of the model in this study were designed to be consistent with land and water resources and crop and livestock production alternatives expected to be possible in 1985.

Table 2.1. Projected per capita commodity demands in 1985[a]

Commodity	Unit	Consumption with 1969 -72 Exports	Consumption with High Exports
Corn	Bushels	1.2010	1.2010
Sorghum	Bushels	0.0486	0.0486
Barley	Bushels	0.5796	0.5796
Oats	Bushels	0.2187	0.2187
Wheat	Bushels	2.5838	2.5838
Oilmeal	Cwt.	-0.0873[b]	-0.0873[b]
Cotton Fibers	Pounds	16.0	16.0
Sugar Beets	tons	0.1089	0.1089
Fed Beef	Lbs. of carcass weight	99.0	74.7
Nonfed Beef	Lbs. of carcass weight	44.6	33.7
Pork	Lbs. of carcass weight	66.7	65.43
Dairy Products	Cwt. of milk equivalent	4.83	4.83
Broilers[c]	Lbs. of ready-to-cook meat	41.1	40.56
Turkeys[c]	Lbs. of ready-to-cook meat	8.6	7.019
Lamb and Mutton[c]	Lbs. of carcass weight	3.1	1.19
Eggs[c]	Number	250.0	250.0

[a]The 1969-72 and high export levels are those shown in Table 3.2

[b]Negative oilmeal consumption reflects an adjustment for the high protein grain by-products provided from the milling of the per capita equivalent of the other grains.

[c]Exogenous activities to the model.

Table 2.2. Projected net export levels for 1985

Commodity	Units	Net Export Level 1969-72[a]	High[b]
Corn	Thou. bu.	626,333	6,800,000
Sorghum	Thou. bu.	126,666	800,000
Barley	Thou. bu.	48,666	400,000
Oats	Thou. bu.	16,179	200,000
Wheat	Thou. bu.	658,719	1,200,000
Oilmeals[c]	Thou. bu.	276,407	350,000
Cotton	Thou. bales	3,306	13,000
Beef	Thou. cwt.	-16,088[d]	-16,088
Pork	Thou. cwt.	-3,349	-3,349
Dairy Products	Thou. cwt.	-4,661	-4,661
Broilers	Thou. cwt.	295,416	295,416
Turkeys	Thou. cwt.	44,162	44,162
Eggs	Thou. doz.	68,699	68,699
Sheep and Lambs	Thou. cwt.	-1,647	-1,647

[a]1969-72 exports are defined as the average of 1969-72 levels.

[b]High exports are defined such that almost the entire agricultural land base of the United States is utilized. As recent U.S. experience has shown, high export levels result in higher commodity prices. Thus, the projections of domestic consumption of meat are reduced because of anticipated consumer response to these price increases. These reductions principally affect beef consumption as per capita demands for fed beef and nonfed beef are reduced by 25 percent in the high export levels compared to the 1969-72 export levels.

[c]Oilmeals are expressed in soybean meal equivalent.

[d]Negative sign indicates imports.

Table 2.3. Levels of allowable soil loss, use of nitrogen and pesticides, and export levels for the alternatives analyzed

Alternative	Allowable Soil Loss Per Acre in Iowa	Nitrogen Limit in Iowa	Pesticide Limit in Iowa	Export Level
A	Unlimited	Unlimited	Unlimited	1969-72 levels
B1	5 tons	Unlimited	Unlimited	1969-72 levels
B2	2.5 tons	Unlimited	Unlimited	1969-72 levels
C	5 tons	100 lbs/acre	Unlimited	1969-72 levels
D	5 tons	100 lbs/acre	Restricted	1969-72 levels
E	5 tons	Unlimited	Unlimited	High exports[a]
F	5 tons	100 lbs/acre	Restricted	High exports

[a]High exports are defined such that the entire agricultural land base of the United States is utilized (see Table 2.2).

ALTERNATIVES ANALYZED

In this study, seven alternatives are analyzed with respect to the application of soil-loss and other environmental restrictions on Iowa agriculture (Table 2.3). Under Alternative A (the baseline alternative), no restrictions on soil loss or chemical usage are applied in Iowa. Alternative B1 assumes the state of Iowa limits the annual soil loss to 5 tons per acre per year. Alternative B2 reduces the allowable soil loss in Iowa to 2.5 tons. Alternatives C and D each incorporate the

5-ton soil-loss limit. Alternative C also limits chemical nitrogen to 100 pounds per acre, and Alternative D, in addition to this 100-pound limitation on nitrogen fertilizer, limits insecticides to only organo-phosphates and carbamates. The restriction on the usage of nitrogen is imposed only on the 12 producing areas in Iowa. Whenever a cropping activity defined in these producing areas in the model uses more than 100 pounds of nitrogen per acre, the use is restricted to 100 pounds and the yields are adjusted accordingly using yield response functions (Nicol and Heady, 1975). The use of insecticides is similarly restricted and the crop yields and production costs are adjusted accordingly. Alternatives E and F are the same as C and D, respectively, except that exports are significantly increased.

RESULTS

The aggregate impact of controls on allowable soil loss and restrictions on the use of nitrogen fertilizer and other chemical inputs under different export levels on Iowa agriculture are shown in Table 2.4. Imposing gross soil-loss limits on Iowa agriculture dramatically reduces soil losses in Iowa. Because the changes in the crop production practices required to attain this soil-loss reduction alter the comparative advantages Iowa has in producing the various crops, interregional shifts in cropping patterns occur. These interregional shifts result in fewer acres cultivated in Iowa, lower land values, and lower net income for the agricultural sector of Iowa. Imposing the additional environmental restraints limiting the use of nitrogen fertilizer and insecticides results in even lower net agricultural income for Iowa and a smaller share of total U.S. net farm income. Importantly, the per capita cost of agricultural products for the nation does not significantly change with the imposition of the soil loss and other environmental controls.

The high export alternatives result in greatly increased net income for Iowa while complying with the soil-loss and other environmental controls. In addition, the high export levels result in significantly higher returns to land in the model, implying substantial increases in the asset values of landowners.

Table 2.4. Values for major variables at the national levels under the various alternatives

Alternative	Soil Erosion in Iowa	Nitrogen Use in Iowa	Land use in Iowa	Cost Per capita	Iowa Farm Income as Percent of U.S. Total Farm Income	Index of Iowa's Net Income
	(000 tons)	(000 tons)	(000 acres)	($)	(%)	(% of A)
A	361,960	711.5	27,209	156.44	10.2	100
B1	107,712	703.5	27,029	156.94	9.9	97.3
B2	60,129	777.0	26,840	157.14	9.6	94.1
C	112,321	637.0	26,664	159.47	9.4	94.7
D	108,810	509.0	26,999	156.98	9.5	93.2
E	108,252	1,449.0	27,919	NA[b]	10.9	263.0
F	112,861	1,083.0	28,024	NA	9.9	260.8

[a]Costs are based on commodity shadow prices determined in the model's solution under each alternative.

[b]NA means not available.

Soil Restrictions in Iowa

Imposing soil restrictions on Iowa farmers to force them to manage the land in less erosive ways results in a loss of farm output from the state as some land is taken out of production. With the imposition of the 5-ton soil-loss limit in Alternative B1 about 178,000 acres of Iowa cropland are taken out of production compared to Alternative A. The data in Table 2.5 show the changes in land utilization in Iowa at a disaggregated level. The lands in groups 1, 2, and 3 are not idled under any of the alternatives. The land in land groups 4, 5, 6, and 7 in the model are those most susceptible to erosion and are of lower productivity. Lands in these groups are idled as the cost per unit of output is increased because of the imposition of the soil-loss restrictions. The owners of these lands in Iowa will realize greatly reduced incomes as their land is forced out of production following their loss in comparative advantage to farmers elsewhere in the nation who are not required to protect their lands from erosion. In addition to the idled cropland, the total acreage of high-value row crops, such as corn and soybeans, in Iowa decreases by 250,000 acres. This decline in row crops is only partially offset by a 27,000-acre increase in close-grown crops.

Under the 5-ton soil-erosion restriction in alternative B1, Iowa farmers significantly reduce the use of conventional tillage in conjunction with straight-row farming compared with Alternative A (Table 2.6). A much larger proportion of the land under conventional tillage is protected with contouring, strip cropping, and

Table 2.5. Cultivated land in Iowa by land groups[a]

Land Groups	Alternative						
	A	B1	B2	C	D	E	F
	(thousand acres)						
1, 2, 3	17,388	17,388	17,388	17,388	17,388	17,388	17,388
4, 6	8,364	8,234	8,060	8,060	8,184	9,035	9,015
5, 7	1,454	1,456	1,389	1,215	1,457	1,504	1,504
8, 9	0	0	0	0	0	11	117

[a]Land groups are as shown in Table 1.1.

Table 2.6. Categorization of the management of Iowa land under the alternatives analyzed

Item	Alternative						
	A	B1	B2	C	D	E	F
Conventional Tillage:							
Straight row	23,232	10,554	4,487	8,647	11,848	6,917	8,553
Contour	215	2,288	4,156	4,499	2,288	353	2,048
Strip crop and terrace	1	4,085	7,098	4,301	4,068	6,687	4,264
Reduced Tillage:							
Straight row	3,064	6,111	4,236	1,993	4,834	7,680	4,192
Contour	0	3,294	3,254	3,171	3,294	3,343	4,988
Strip crop and terrace	697	697	3,609	4,053	667	1,939	3,979

terracing in Alternative B1. There is also a significant shift from conventional tillage to reduced tillage. Following these shifts in production patterns and farming practices, soil-erosion losses from Iowa cropland decline from 362 million tons under Alternative A to less than 108 million tons under Alternative B1.

Under Alternative B2, in which the soil-loss limit in the model is reduced to 2.5 tons, further changes occur in Iowa production patterns and farming practices. Gross soil loss from Iowa lands decreases to 60 million tons. (This reduced soil loss is more than offset by increases elsewhere in the country as crop production shifts away from Iowa.) The net result of the interregional shifts is a further reduction of 357,000 acres of cropped land in Iowa compared to Alternative B1.

Reducing the allowable soil loss from 5 tons to 2.5 tons per acre results in an even greater use of contouring, strip cropping, and terracing to protect the cropland under conventional tillage (Table 2.6). In addition, a small proportion of the land tilled by conventional tillage methods is shifted to reduced tillage under Alternative B2 compared to Alternative B1. Comparing both B1 and B2 alternatives to the unrestricted Alternative A shows the very significant shift to the use of strip cropping and terracing from straight-row farming required to meet these soil-loss limits.

Total gross soil erosion from Iowa land in land groups 1, 2, and 3 of the model decreases from 122 million tons in Alternative A to 75 million tons in Alternative B1 to less than 39 million tons in Alternative B2. This reduction is due mainly to the increased use of reduced tillage and the shift from straight-row cropping to contouring, strip cropping, and terracing. A similar change in farming practices occurs on the more erosive lands in land groups 4 and 5. The area in reduced tillage increases from .8 million acres in Alternative A to 2 million acres under Alternative B1 to 3 million acres in Alternative B2. This shift away from conventional tillage, especially conventional tillage in conjunction with straight rows, sharply reduces the gross soil loss on the lands in groups 4 and 5 relative to the erosion from land groups 1 and 2. In Alternative A about 33 percent of the total gross soil erosion in Iowa is from land groups 1 and 2. The excessive erosion from the more erosive lands is reduced so substantially in Alternative B2 that, with taking some of these erosive lands out of production, the relative contribution of land groups 1, 2, and 3 increases to 62 percent of the total soil eroded in Alternative B2.

Environmental Restrictions in Iowa

Imposing additional environmental restrictions on Iowa agriculture under Alternatives C and D reduces both Iowa's net farm income and Iowa's share of total U.S. net farm income compared to Alternative B1. (Alternatives B1, C, and D all incorporate a 5-ton soil-loss limit.) The total cropped acreage in Iowa decreases by 300,000 acres in Alternative C compared to Alternative B1. Crop yields in Iowa decrease only slightly in the model under Alternative C, as changes in farming practices partly offset the nitrogen limitation. More corn is grown in rotation with legumes and under better farming practices. Under Alternative D, the use of both nitrogen fertilizer and insecticides is limited. Because of the resulting yield reductions, more acres are brought into production under Alternative D than Alternative C. The area in crops in Iowa increases from 32.0 million acres under Alternative C to 32.4 million acres under Alternative D as land is substituted for these restricted inputs.

Effects of High Export Levels in Iowa

The export levels in the alternatives reviewed above are the average of the export levels in the years 1969-72. With these levels, U.S. agriculture had surplus capacity. Two alternatives of the model are specified with export levels designed to almost exhaust the productive capacity of U.S. agriculture. The first high-export alternative, Alternative E, has a 5-ton soil-loss limit for Iowa. The second, Alternative F, adds the environmental restrictions on the use of nitrogen and pesticides in conjunction with the 5-ton soil-loss limit.

With the high export levels under Alternatives E and F, the nation, including Iowa, is required to increase the land area under cultivation (Table 2.4). For Iowa this requires bringing back into production the land in land groups 4, 5, 6, and 7 that the restrictions idled in Alternatives B1, C, and D. (See Table 2.5.) Large amounts of additional nitrogen fertilizer are also required in Iowa (as well as elsewhere) to meet the increased commodity demands. Use of nitrogen under Alternative E is more than twice that under Alternative A, where the nation's agriculture has surplus capacity. Under Alternative F, the use of nitrogen and insecticides in Iowa is restricted, resulting in a substitution of land for these inputs. The total quantity of nitrogen used decreases by 25 percent in Alternative F compared to Alternative E. The land area cropped increases by 105,000 acres from E to F.

Farm Prices and Iowa Farm Income

The five alternatives (A, B1, B2, C, and D) show little impact on commodity prices and farm income at the national level (Tables 2.7 and 2.8). The crucial variable with respect to prices and income in the study is the export level. Under Alternatives A, B1, B2, C, and D, prices for corn, wheat, and soybeans are at modest

Table 2.7. Average prices for selected commodities under each alternative[a]

	Alternative						
Commodity	A	B1	B2	C	D	E	F
	(dollars/unit)						
Corn (bu.)	1.11	1.10	1.08	1.12	1.12	2.51	2.73
Wheat (bu.)	1.17	1.16	1.14	1.16	1.16	3.02	3.13
Soybeans (bu.)	3.70	3.66	3.54	3.77	3.67	9.53	8.60

[a]These prices are calculated from a weighted average of the prices obtained in each of the market regions. The prices are in 1970 dollars.

Table 2.8. Comparison of crop production costs and net farm income at the state and national level under the various alternatives

Item	Alternative						
	A	B1	B2	C	D	E	F
Iowa:							
Crop costs	100	105	108	104	104	139	123
Net income	100	97	94	95	93	263	261
Remainder of U.S. Agriculture:							
Crop costs	100	99	99	100	99	145	146
Net income	100	100	100	104	101	245	271

levels. Under the high export levels of Alternatives E and F, commodity pricese are much higher.

Imposition of soil-loss limits makes farming in Iowa less profitable relative to the rest of the country. Total crop production costs in Iowa increase from $1.7 billion (Alternative A) to $1.8 billion under the 2.5-ton soil-loss restriction (Alternative B2). As expected, net farm income in Iowa decreases with the imposition of soil-loss restrictions from $2.0 billion under Alternative A, with no soil-loss restrictions, to $1.9 billion with the imposition of a 2.5-ton restriction. At the same time, income increases in the rest of the country from $17.8 billion to $17.9 billion.

The higher export level has a substantial impact on Iowa farm income. Net farm income in Iowa increases from $2.0 million in Alternative A to $5.3 billion in Alternative C, a 163 percent change, due to the increase in exports. At the same time, net farm income in the rest of the country increases 145 percent to $43.5 billion. In other words, Iowa still retains a substantial comparative advantage within the U.S. agricultural sector in producing the additional grain needed to meet the increased grain export levels, even with a 5-ton soil-loss restriction.

This advantage is partially lost with the addition of the environmental restrictions for nitrogen fertilizer and insecticides. With these additional restrictions, farm income in Iowa declines slightly in Alternative E compared to Alternative F. This loss in Iowa farm income is a gain for the rest of the country as the net income from the remainder of the agricultural sector increases from $43.6 billion to $48.1 billion when Alternatives E and F are compared.

The increased exports require the cultivation of some of the less productive lands in Iowa and the rest of the country. This utilization of land in Alternatives E and F that is idle in Alternatives A and B1 is reflected in the higher commodity prices and the increase of land rental values (shadow prices). The rental price of land increases by 450 percent under Alternatives E and F compared to Alternative A.

CONCLUSION

The farmers in Iowa will bear the cost of environmental improvement. This burden will fall mostly on those landowners owning lands that are both susceptible to erosion and of low productivity. Without the pressure on the agricultural sector from high export demands, such land will be forced out of production. Benefits accrue to the nation's consumers generally in improved environmental quality and to farmers elsewhere in the form of higher incomes. An income redistribution will take place, with farmers in Iowa sacrificing and consumers and other farmers realizing the benefits.

The sacrifice made by the state's farmers might be handled in different ways: First, society could decide that farmers should bear the economic cost because they own the land. Second, the state could pay an annual subsidy to the farmer equal to the amount of income he sacrifices as he shifts land use and farming systems to conform with land-use and environmental regulations. Third, the environmental program could be applied on a national basis.

In Chapter 9, we employ a hybrid econometric and recursive programming model to analyze Iowa conservation alternatives.

REFERENCES

Iowa State University Cooperative Extension Service. 1972. Iowa's New Conservancy Districts and Soil Loss Limit Regulations. PM-536. Ames.

Nagadevara, V. S. S. V., E. O. Heady, and K. J. Nicol. 1975. Implications of Application of Soil Conservancy and Environmental Regulations in Iowa in a National Framework. CARD Report 57. Center for Agricultural and Rural Development, Iowa State University, Ames.

U.S. Department of Agriculture, Conservation Needs Inventory Committee. 1971. National Inventory of Soil and Water Conservation Needs 1967. Statistical Bulletin 461. Washington, D.C.

U.S. Water Resources Council. 1970. Water Resources: Regions and Subregions for the National Assessment of Water and Related Land Resources. U.S. Government Printing Office, Washington, D.C.

SECTION II. ADEQUACY OF LAND AND WATER RESOURCES FOR U.S. AGRICULTURE UNDER ENVIRONMENTAL IMPROVEMENT PROGRAMS

THE national interregional competition studies in this section are formulated for studying the relationships of agricultural production to land and water use and the environment. Particular emphasis is given to identifying national and regional resources that are likely to be in critical supply in the future if agricultural practices are modified by government policy to alter agriculture's impact on the environment. The first chapter in this section evaluates the adequacy of the nation's land and water resources under varying assumptions concerning available technology and national resource policy. Emphasis is given to future water supplies for irrigation in the western United States. The second study utilizes a modified model from the first study to evaluate the impact of national policies designed to curb pollution problems caused by excessive erosion of the soil, persistence of certain organochlorine insecticides in the environment, feedlot runoff, and nitrogen fertilizer use. In the third chapter the same model is used to analyze tradeoffs between national goals for U.S. agriculture of production efficiency and soil-loss reduction.

The analysis in the first study indicates that national resource policies aimed at protecting the environment, especially policies affecting future water supplies for agriculture, may be in conflict with national policies aimed at increasing agricultural output for export. The evaluation of the environmental policies in the second study emphasizes the resource substitutions that occur when the use of certain agricultural practices is prohibited or restricted. The results provide an indication of the potential flexibility of varying aggregate resource use over the long run as an adjustment to the policies investigated. In the third study, the tradeoff analysis between the national goals of minimizing the cost of producing food and fiber and reducing soil erosion losses shows that at high levels of soil loss a given reduction in erosion can be obtained without substantial cost to society. When soil losses are at relatively low levels, however, further reductions are very expensive.

CHAPTER 3. ASSESSMENT OF WATER AND LAND RESOURCES FOR U.S. AGRICULTURE WITHIN AN INTERREGIONAL COMPETITION FRAMEWORK

by Anton D. Meister and Earl O. Heady

THIS STUDY was carried out in conjunction with the U.S. Department of Agriculture (USDA), Economic Research Service (ERS), and the Soil Conservation Service (SCS) as a part of the 1975 National Water Assessment. The program was under the auspices of the nation's Water Resource Council. The main objectives of the study were to evaluate the nation's land and water resource capabilities for agriculture under varying assumptions of technology and resource policy. Production of food affects the environment in both beneficial and adverse ways. If the effect is adverse, society may curtail or modify certain agricultural production activities through policies or laws to maintain or improve the environment. The adequacies of land and water resources to meet future agricultural demands may then have to be reconsidered. In this study on land and water adequacy, emphasis is on the nation's water resources.

The increasing levels of domestic and export demand of recent years have been met in two ways: more land has been brought into production, and existing cropland has been cropped more intensively. The environmental problems associated with the use of more land are primarily those of wind and water erosion and the loss of wildlife habitat (for example, through drainage of wetlands). The problems with the more intensive use of land result primarily from heavy use of fertilizers and pesticides and increasing soil and water salinity because of irrigation.

Concern about water erosion and sediment has been one of the basic elements of the conservation movement during the past four decades. But within that period a significant change in emphasis has taken place. The soil conservation efforts that originated in the 1930s were first concerned with physical destruction or waste of soils resulting from man-induced erosion. The intrinsic costs to present and future generations were expressed as reduced capacity for agricultural production, increased flood hazards, and adverse social and economic effects for landowners, communities, the states, and the entire nation (Weinberger and Hill, 1971).

In years past, water development projects and water-related activities at both state and federal levels often went forward with little regard for damage caused to fish and wildlife resources. Thousands of miles of natural stream channels were relocated or altered, some streams were dried up, estuaries and marshes suffered from drainage and landfill operations, and estuarine habitat essential for shellfish and other species was destroyed by dredging and channel deepening. Water quality deterioration has also adversely affected fish and wildlife resources in both marine and fresh water (National Water Commission, 1973). A rising level of population and affluence will, among other things, increase the demand for food and the demand for recreational areas with game for hunting and fishing. In terms of land and water resources development, these can be conflicting demands.

Another major pollutant in agriculture is animal wastes. Whereas in the past only hundreds of cattle were fattened or poultry housed, now thousands can be accommodated in the same area at the same time. This concentration has led to waste disposal problems. Runoff from feedlots and other large-scale livestock production units has become an increasing source of pollution of the rivers and lakes.

The aim of this study is to evaluate the magnitude of some of the tradeoffs between agricultural output and environmental quality using a model of the nation's agricultural sector. For example, by restricting wetland or forestland development in this model, it is possible to determine how necessary this type of development really is to meet future demands for agricultural products. A summary of this analysis of restricting land development, as well as other policy measures to improve environmental quality, is presented in this chapter. A description of the model, the organization of this study, and a report of principal results precedes the analysis summary.

DESCRIPTION OF THE MODEL

This study uses a linear programming model for the nation's agricultural sector similar to the model in Chapter 1. However, this model encompasses 105 producing areas (Figure 3.1) instead of the 223 in Chapter 1 because this number of producing areas still provides sufficient data at national and regional levels. The 9 land groups defined in Table 1.1 are retained in each of the 105 producing regions. Water regions are increased to 58 (Figure 3.1) to provide more detail on this resource. Finally, market regions are reduced from the 30 in Figure 1.3 to the 28 in Figure 3.2. Producing areas and water regions again are contained in the different market regions, and the market regions are interconnected by a transportation sector. This set of regions, together with the transportation network, make the model capable of analyzing the major effects of proposed environmental restrictions and other changes in policy parameters. The interregional linkages simulate the dependence that exists among the different areas and activities in the agricultural sector. A restriction on the use of water for irrigation in the West (because of higher demands for municipal and industrial uses or because of minimum stream flow restrictions for fish and wildlife) will affect the level of production in the Midwest and the East. Similarly, restrictions on soil loss will move production out of soil-loss prone areas into other areas. It is this interdependence that makes the model a suitable analytical device for studying regional shifts in production and land-use patterns resulting from changes in regional comparative advantage caused by policy changes.

The model again is a partial equilibrium model that, given assumptions about levels of consumption and exports, minimizes the cost of producing this quantity

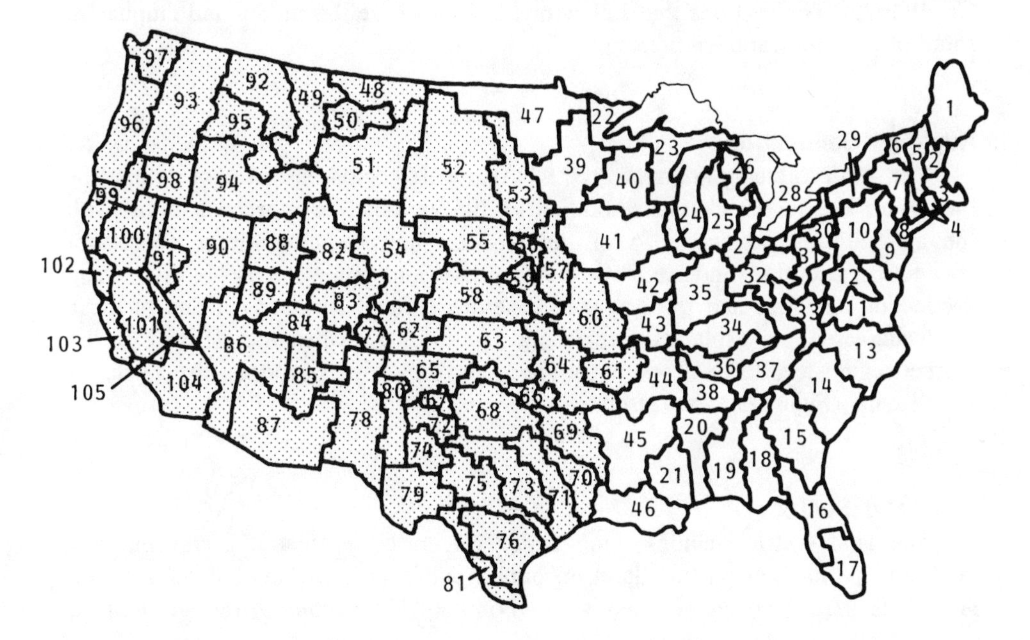

Figure 3.1. The 105 producing areas (areas with irrigation shaded).

Figure 3.2. Twenty-eight market regions (MR) of the United States.

of agricultural product demand within the restraints imposed on the model. The model assumes all farm resources receive their market rate of return (except for land and water, where return is determined endogenously in the model).

A complete and detailed description of the model is given in Meister and Nicol (1975). The sections that follow briefly summarize the model and emphasize some of its important features.

Regional Delineations

The 105 producing areas are the 99 Aggregated Subareas (ASAs) defined by the Water Resources Council modified to 105 areas to be more consistent with the agricultural patterns in 6 of the ASAs. The producing areas consist of contiguous counties and sum to both ASAs and major river basins. Crop production activities and the cropland base are defined within each one of these producing areas as explained previously. Livestock activities are defined at the market-region level. For economy of space and reporting purposes, the 18 major river basins (Figure 3.3) are used.

The Land Sector

The land sector includes three major categories: cropland, permanent hay land, and permanent pasture land, including public grazing lands and forestland grazed. The cropland sector again is based on the *1967 National Inventory* (USDA,

Figure 3.3. River basins with county boundaries.

Conservation Needs Inventory Committee, 1971) cropland definition with an adjustment for wild hay as determined from the *1969 Agricultural Census* (U.S. Department of Commerce, Bureau of Census, 1971). The remaining lands, which are incorporated in the model as available hay equivalents to provide a base level of pasture, are determined by data from the Conservation Needs Inventory (CNI) pasture, range, and forestland grazed, with the additional public lands grazed determined from the Census.

The cropland base from the *National Inventory* is aggregated from the county level to the 105 producing areas by the 9 land groups shown in Table 1.1. This land aggregation is adjusted to 1974 actual acreage by adding lands developed through drainage conversion from the Class IIw and IIIw wet soils to a Class I equivalent productive capacity. Adjustments in irrigated acres are made to reflect developments after 1967.

Acreages to be irrigated from future public development projects are also included in the irrigated cropland base. Projected acreages to be included in the irrigated cropland base from future public irrigation development projects are assumed to be 85 percent of the full-service acreage in authorized and funded projects that are to be in place by 1985 and 2000 (Table 3.1). These acreages,

Table 3.1. Potential public and private irrigated land development by producing areas

Producing area	Private 1975 through		Public 1975 through	
	1985	2000	1985	2000
	(acres)			
51	15,148	33,844	0	0
52	27,898	61,215	0	6,520
53	5,004	11,104	84,120	244,050
54	128,385	286,327	6,980	69,980
55	405,824	679,543	230,000	270,000
56	20,646	45,876	0	0
57	2,581	5,934	0	0
58	134,446	300,452	0	0
59	174,568	387,662	0	0
60	470	1,042	0	0
61	-	-	0	0
62	8,020	18,164	0	0
63	389,384	864,018	0	0
64	8,072	17,889	0	0
65	75,510	136,826	0	0
66	7,556	16,303	0	0
67	34,723	34,723	0	0
68	109,522	229,133	0	0
69	3,204	6,662	0	0
70	5,011	5,011	0	
71	14,319	14,319	0	0
72	155,880	165,230	0	0
73	8,233	8,233	0	0
74	30,136	34,575	0	0
75	15,392	15,392	0	0
76	16,108	16,108	0	0
77	9,150	20,723	0	0
78	17,476	32,208	0	0
79	8,949	8,949	0	0

Table 3.1. (continued)

Producing area	Private 1975 through 1985	Private 1975 through 2000	Public 1975 through 1985	Public 1975 through 2000
	(acres)			
80	8,885	19,698	0	0
81	38,303	38,303	0	0
82	12,366	27,692	52,760	52,760
83	10,110	22,897	58,520	58,520
84	4,164	9,364	85,795	180,795
85	903	2,037	0	0
86	8,038	18,098	0	0
87	35,252	79,597	0	0
88	19,795	24,053	17,500	17,500
89	-	-	26,240	26,240
90	23,653	51,987	0	0
91	15,092	33,170	0	0
92	1,450	3,236	3,600	3,600
93	45,020	100,555	62,400	256,700
94	209,937	467,785	49,000	49,000
95	14,487	322,295	0	0
96	4,374	9,764	11,200	11,200
97	1,927	4,304	0	0
98	3,392	7,568	0	0
99	4,597	10,311	0	0
100	13,252	30,121	0	0
101	35,382	80,422	0	0
102	2,807	6,380	14,700	14,700
103	2,872	6,528	0	0
104	9,335	21,219	0	0
105	326	742	0	0
U.S.	224,9386	4,567,923	74,5465	1,392,235

from future private irrigation development as projected in the ERS Series E' projections, are defined in the model as available for incorporation into the model's irrigated cropland base through the economic decision processes of the model (Meister and Nicol, 1975). An irrigation development cost is calculated for each ASA with dryland available for conversion under private irrigation development. When the model is solved, land is converted for irrigated crop production only if the crops grown on this converted land have a lower cost, including the conversion costs, than elsewhere on the model's dryland and irrigated cropland base.

Projections are made of the conversion of Class IIw and IIIw pastureland and forestland to cropland for the years 1985 and 2000. Cropland conversions projected after 1973 are not included in the cropland base but represent a supply of land that may be converted to nonirrigated crop use in the solution of the model (Table 3.2). Conversion cost data for clearing and draining these wetlands are incorporated in the model as conversion costs to transform these noncropland wetlands to cropland use. When the model is solved, land is converted for crop production only if the crops grown on this converted land have a lower cost, including the clearing and draining costs, than elsewhere on the model's dryland and irrigated cropland base.

Table 3.2. Potential wetland development for 1985 and 2000

Producing area	Maximum Pasture Conversion 1985	Maximum Pasture Conversion 2000	Maximum Forest Conversion 1985	Maximum Forest Conversion 2000
	(thousand acres)			
6	13.6	32.1	16.9	40.0
11	32.7	77.3	173.3	409.7
13	36.5	86.3	358.6	911.4
14	74.8	176.7	395.1	933.8
15	70.1	165.7	288.9	682.6
16	141.4	334.2	258.1	610.1
18	71.7	169.6	193.4	457.1
19	100.7	238.0	182.0	430.2
20	116.6	275.7	123.4	291.7
21	94.5	223.3	162.5	384.0
22	9.9	23.3	187.4	442.8
25	68.1	160.9	157.9	373.2
27	44.2	104.4	75.9	179.4
29	40.5	95.7	121.3	286.7
31	42.7	101.0	31.0	73.2
32	78.9	186.4	39.8	94.2
34	48.6	115.0	42.9	101.5
35	130.2	307.7	76.8	181.4
36	21.7	51.4	16.4	38.7
37	24.6	58.1	9.4	22.3
38	54.9	129.8	43.4	102.6
39	151.7	358.6	156.9	370.9
40	142.1	335.9	158.6	374.8
41	203.9	482.1	27.8	65.8
42	54.6	129.1	24.3	57.5
43	50.2	118.8	38.5	91.1
44	52.1	123.2	150.9	356.7
45	191.4	452.5	322.1	356.7
46	108.6	256.7	180.4	308.2
47	89.8	212.2	105.9	250.4
53	130.5	308.5	4.8	11.4
55	15.0	35.5	3.3	7.9
56	18.6	43.9	1.4	3.3
57	63.1	149.1	8.5	20.1
60	116.4	275.2	62.6	147.9
61	30.1	71.1	23.1	54.6
64	144.8	342.3	57.7	136.3
69	158.4	374.4	114.3	270.2
70	122.3	289.0	138.3	326.9
U.S.	3160.5	7470.7	4560.8	10662.0

Source: U.S. Department of Commerce, Bureau of Census, 1971.

The land base is adjusted for expected conversion to urban and other nonagricultural uses between 1967 and the target dates of the analysis, 1985 and 2000. Adjustments in the land base are also made for crops not endogenous to the model. Excluded from the model are the more intensely grown crops mentioned in Chapter 1. For these crops, irrigated and nonirrigated acreage projections from the USDA's National Interregional Agricultural Projections Systems (NIRAP) model are obtained for each ASA. Projected exogenous crop acreages are subtracted from the total cropland and irrigated bases prior to solution of the model. This prior adjustment is justified on the basis that these crops are generally the higher-value crops and would have a relative economic advantage in competition for land use.

The Water Sector

The water sector is only defined west of the Mississippi, except the Souris-Red-Rainy-Basin, as shown in Figure 1.2. The water supply in each water supply region is derived in a manner consistent with the estimating procedure as outlined in the Volumetric Adequacy work statement of the Water Resources Council (1974b). Producing-area water supplies are calculated at present-use levels plus seven-tenths of the outflow from the region during the months when irrigation of water is applied to crops.

The consumptive use of water for irrigation in 1975 was calculated by multiplying the 50 percent precipitation coefficients for irrigation by the figure for 1971-73 average irrigated acres reported by the Statistical Reporting Services (SRS), or the figure for NIRAP projected areas for crops not reported by SRS. These water-use estimates thus include irrigation from both surface and ground water. Depletion of ground water resources over time is accounted for in the final water supply available in the year 2000.

Water-use coefficients for crop activities in the model represent the net diversion requirement to provide the crop with the amount of water needed for growth in addition to that provided from precipitation. The irrigation requirements of individual crops are based on current irrigation delivery and application practices with assumed rainfall at the 80 percent exceedance level. (This represents a relatively dry year since more rainfall would be expected to occur in eight out of ten years.) Two sets of water-use coefficients were obtained from the Soil Conservation Service staff in Denver for this use. The first set represented projected water-use coefficients based on trend increases in water-use efficiency. The second set represented high-efficiency coefficients reflecting accelerated changes in delivery and application systems. A water transportation network is developed that reflects natural flows and interbasin transfers. Water prices are acreage-weighted, average, reimbursable costs of the Bureau of Reclamation projects (Meister and Nicol, 1975).

The supplies of water are adjusted for water use by the exogenous crops, irrigated hay land, and rangeland prior to solving the model. However, the model gives the choice to return the irrigated exogenous pasture back to dryland pasture if this is economical. A change of irrigation pasture back to dryland pasture makes irrigation water available for use in other cropping activities. Hay yield of the exogenous pasture is adjusted accordingly if this activity takes place in the model.

The Crop Sector

The crop sector represents the production of barley, corn, corn silage, cotton, legume hays, nonlegume hays, other hays, oats, cropland pasture, other pasture, sorghum, sorghum silage, soybeans, sugar beets, and wheat. These crops are combined in relevant rotations to provide a range of crop production possibilities to be evaluated under the various alternatives. The rotations are combined with the twelve possible conservation and tillage practices mentioned in Chapter 1. The tillage alternatives are residue removal (generally fall plowing), residue management, and conservation or reduced tillage. The three tillage methods are combined with the four conservation alternatives (straight-row cultivation, contouring, strip cropping, or terracing) to provide a large variation in crop-production and soil-loss possibilities (Table 3.3).

Table 3.3. Allowable conservation practices on the different land groups[a]

Land Groups	Row Cropping	Contouring	Strip Cropping	Terracing
1	X[b]			
2	X	X	X	X
3	X	X		
4	X		X	X
5	X	X		
6	X		X	X
7	X	X		
8	X			
9	X			

[a]Based on recommendation from the Soil Conservation Service.

[b]X = Practice allowed.

Gross soil loss, as calculated here, represents the average annual tons of soil leaving the field. This measurement of soil loss does not represent the amount reaching the stream or bodies of water. Some soil particles settle out or are diverted as the runoff passes through grassed areas or onto flatter terrain. Again, two procedures were used to determine the gross soil loss per acre: for the areas east of the Rocky Mountains the Universal Soil-Loss Equation was used; for areas west of the Rocky Mountains data from the Soil Conservation Service questionnaire were used.

As in previous models and throughout other studies in this book, the remaining exogenous crops are accounted for prior to model construction and adjustments are made in the land and water availability. Within the producing regions crop acreages are restrained to be between prespecified upper and lower limits. The crops involved are wheat, corn, silage (corn silage and sorghum silage), sorghum, soybeans, cotton, and sugar beets. The reason for these adjustment constraints is that regional shifts in production are gradual, not instantaneous, due to imperfect mobility of resources. Regional adjustment constraints are based on the crop production patterns reported in the *1969 Census of Agriculture* (U.S. Department of Commerce, Bureau of Census, 1971). Acreages of individual crops are allowed to decrease to 70 percent of the 1969 census acreage by 1985 and to 40 percent by the year 2000. Acreages of crops are allowed to double from 1969 to 1985; no upper limit is set on year 2000 acreages.

The other endogenous crops (oats, barley, and nonlegume hay) are constrained at the market-region level. These market-region crop-acreage restraints are only lower limits and are calculated in the same way as the producing-area adjustment restraints. The model also has a restraint that controls the ratio between the acres of the legume and the nonlegume hays.

The Endogenous Livestock Sector

This sector includes the beef cow, dairy, hog, and beef feedlot activities for producing the following livestock commodities: milk, fed beef, cull beef, pork, and feeders (an intermediate commodity used in the beef feedlot activities). The livestock activities are defined with rations that have previously been developed (not determined endogenously). The activities defined for feeders allow for 16 alternative rations, for hogs and beef cows 5 each are allowed, and for dairy there are 6. Linear combinations of these activities for any livestock type provide an even larger number of possible rations for the livestock.

Regional livestock activity levels are restrained for reasons similar to the crop restraints. The 1985 restraints are set at 80 percent of 1969 for the lower limit and 250 percent of 1969 as the upper limit. The lower restraint level for the year 2000 is set at 60 percent of 1969. All 1969 levels are determined from the *Census of Agriculture* livestock totals.

Livestock wastes historically have served as a ready source of plant nutrients. In line with the restrictions on animal waste runoff into the nation's waterways, all livestock activities considered in the model are developed with the possibility that their wastes, using the "conventional" system of handling, can be utilized as a fertilizer in the cropping sector.

The Demand Sector

Domestic and net export demands are estimated for each commodity based on projected per capita income, consumption, and relative future prices. Exports are based on estimated international prices, expected governmental international agreements, and other demand factors estimated in the OBERS projections. (OBERS stands for the *O*ffice of *B*usiness *E*conomics, now Bureau of Economic Analysis, and *E*conomic *R*esearch *S*ervice. The projections were prepared for the U.S. Water Resources Council.) Total demands include feed grains for exogenous livestock. Projected per capita demand levels and net export levels are shown in Tables 3.4 and 3.5.

The Transportation Sector

Interregional interdependence is created by national and regional demands and by a transportation network. The transportation network is based on the 28 market regions and transports the commodities barley, corn, oats, sorghum, oilmeals, wheat, milk, fed and nonfed beef, pork, and feeders among market regions. Silages, hays, and livestock wastes are transported only within market regions.

Table 3.4. **Projected OBERS E' per capita demand levels for 1985 and 2000**

Commodity	Unit	OBERS E' Consumption 1985[a]	2000[b]
Barley	bushels	0.042	0.05
Corn	bushels	1.207	1.309
Oats	bushels	0.212	0.212
Sorghum	bushels	0.0	0.0
Wheat	bushels	2.472	2.338
Soybeans	bushels	NA[c]	NA[c]
Cotton	bales	0.029	0.025
Beef & veal (carcass wt.)	Pounds	136.7	150.7
Milk (fresh equiv.)	Pounds	511.4	456.6
Pork (carcass wt.)	Pounds	68.1	71.5
Lamb & mutton (carcass wt.)	Pounds	1.8	1.7
Turkeys (R.T.C.)[d]	Pounds	10.9	12.8
Chickens (R.T.C.)	Pounds	49.8	56.5
Eggs	dozen	25.0	38.0

[a]Projected U.S. population for 1985 is 233.1 million.

[b]Projected U.S. population for 2000 is 262.4 million

[c]Not available

[d]Ready-to-cook

Source: U.S. Water Resources Council, 1975.

Table 3.5. Projected OBERS E' net export levels for 1985 and 2000

Commodity	Unit	OBERS-E' 1985	OBERS-E' 2000	OBERS-E' (High) 1985	OBERS-E' (High) 2000
		(thousand units)			
Barley	bushels	20	35	25	40
Corn	bushels	989	2,069	1,889	3,209
Oats	bushels	10	21	19	29
Sorghum	bushels	160	380	270	450
Wheat	bushels	774	919	1,179	1,479
Soybeans	bushels	950	1,475	1,125	1,700
Cotton	bales	4.1	4.2	4.2	4.6
Beef & veal (carcass wt.)	pounds	-2,169	-2,924	-1,190	-1,760
Milk (fresh equiv.)	pounds	-680	-1,040	-680	-1,040
Pork (carcass wt.)	pounds	-307	-351	-307	-351
Lamb & mutton (carcass wt.)	pounds	-230	-274	-230	-275
Turkeys (R.T.C.)[a]	pounds	70	80	70	80
Broilers (R.T.C.)	pounds	235	253	235	253
Eggs	dozen	44	50	43.9	50

[a]Ready-to-cook

Source: U.S. Water Resources Council, 1975.

A MATHEMATICAL EXPLANATION OF THE MODEL

The linear programming model for this particular analysis forms a simultaneous equation system explained below. Its specification was developed by Iowa State University personnel and an interagency group representing the Water Resource Council of the U.S. federal government.

The Objective Function

The objective function is defined to minimize the cost of producing the given agricultural commodities subject to the restraints on the availability of land, water, fertilizer, amount of crop and livestock adjustment allowed, and the intermediate commodities demanded. It can be represented by:

$$\begin{aligned} \min F = & \sum_i \sum_j \sum_k \sum_m X_{ijkm} XC_{ijkm} + \sum_n \sum_p \sum_q LC_{npq} \\ & + W_r WC_r + F_n FC_n + IB_r IC_r + \sum_n \sum_s \sum_t TC_{nst} \\ & + \sum_i (LD_i + DC_i + RD_i RC_i) \end{aligned} \tag{3.1}$$

i = 1, ..., 105 for the producing areas;
j = 1, ..., 18 for the land groups;
k = 1, ..., 330 for the rotations defined;
m = 1, ..., 12 for the conservation and tillage alternatives;
n = 1, ..., 28 for the market regions;
p = 1, ..., 4 for the endogenous livestock classes;
q = 1, ..., 32 for the livestock rations;
r = 1, ..., 58 for the water supply regions;
s = 1, 2, 4, 5, 7, 8, 10, 11, 13, 14, 15 for the commodities transported;
t = 1, ..., 176 for the transportation routes defined.

Equations of the System

As with earlier models, this model has a set of equations for each producing area, water supply region, and market region. Also, equations at the national level are specified for cotton and sugar beets. To summarize these equations, a typical producing area with irrigation (those without irrigation have equations deleted which relate to water supplies and irrigated crops), a water supply region, and a commodity market region will be described.

Producing area. Each producing area has restraints for land availability by the 9 dry and irrigated land groups, flexibility restraints to control the level of production of 8 crops, and a restraint to define a minimum irrigated acreage. The equations for the ith producing area are as follows:

Dryland restraint by land group:

$$\sum_k \sum_m X_{ikm} AD_{ijkm} + LD_i LDP_{ij} - RD_i RDP_{ij} \leq DA_{ij} \tag{3.2}$$

i = 1, ..., 105 for the producing areas;
j = 1, ..., 9 for the land groups;
k = 1, ..., 330 for the rotations defined;
m = 1, ..., 12 for the conservation-tillage alternatives.

Irrigated land restraint by land class:

$$\sum_k \sum_m X_{ijkm} AI_{ijkm} + RD_i RDP_{ij} \leq IA_{ij} \quad (3.3)$$

i = 48, ..., 105 for the producing areas;
j = 1, ..., 18 for the land groups;
k = 1, ..., 330 for the rotations defined;
m = 1, ..., 12 for the conservation-tillage alternatives.

Crop acreage restraint:

$$MINA_{iu} \leq \sum_j \sum_k \sum_m X_{ijkm} W_{ijkmu} \leq MAXA_{iu} \quad (3.4)$$

i = 1, ..., 105 for the producing areas;
j = 1, ..., 18 for the land groups;
k = 1, ..., 330 for the rotations defined;
m = 1, ..., 12 for the conservation-tillage alternatives;
u = 2, 4, 11, 13, 14, 15 for the crops restrained at the producing-area level for corn, silage, cotton, sorghum, soybeans, sugar beets, and wheat.

Hay acreage restraint:

$$\sum_j \sum_k \sum_m X_{ijkm} W_{ijkm5} \leq HR_i\left(\sum_j \sum_k \sum_m X_{ijkm} W_{ijkm6} + \sum_j \sum_k \sum_m X_{ijkm} W_{ijkm5}\right) \quad (3.5)$$

i = 1, ..., 105 for the producing areas;
j = 1, ..., 18 for the land groups;
k = 1, ..., 330 for the rotations defined;
m = 1, ..., 12 for the conservation-tillage alternatives.

Irrigated acreage restraint:

$$\sum_j \sum_k \sum_m X_{ijkm} AI_{ijkm} \geq \sum_u AIC_{iu} \quad (3.6)$$

i = 48, ..., 105 for the producing areas;
j = 9, ..., 18 for the land groups;
k = 1, ..., 330 for the rotations defined;
m = 1, ..., 12 for the conservation-tillage alternatives;
u = 1, ..., 6, 8, 11, ..., 15 for the crops irrigated.

In the producing areas 48-105, water supplies and irrigation activities are defined. The following equation controls the allocation of water to the endogenously determined agricultural uses:

$$\sum_j \sum_k \sum_m \sum_u X_{ijkm} W_{ijkmu} CWU_{iu} + \sum_n \sum_p \sum_q Y_{npq} LWU_{npq} LW_{npr} - WH_r WA_r \leq WS_r \tag{3.7}$$

i = 48, ..., 105 for the producing areas;
j = 1, ..., 18 for the land groups;
k = 1, ..., 330 for the rotations defined;
m = 1, ..., 12 for the conservation-tillage alternatives;
n = 1, ..., 28 for the market regions;
p = 1, ..., 4 for the endogenous livestock types;
q = 1, ..., 32 for the livestock rations;
r = i, ..., 47 to give the water supply region number;
u = 1, ..., 15 for the possible irrigated crops.

Commodity market regions (Figure 3.2). To reflect the commodity requirements based on domestic demand and foreign trade movements through the region, each commodity market region has a set of equations to balance the supply and demand of the commodities. Each region also has a set of equations to equate the level of livestock production with livestock demand in the region.

The equations are

Commodity balance equation:

$$\sum_i \sum_j \sum_k \sum_m X_{ijkmn} W_{ijkmu} CY_{ijkmsu} + \sum_p \sum_q Y_{npq} LY_{npqs} - \sum_t T_{nst} - \sum_r WH_r DA_{rs} \geq CD_{ns} \tag{3.8}$$

i = 1, ..., 105 for the producing areas;
j = 1, ..., 18 for the land groups;
k = 1, ..., 330 for the rotations;
m = 1, ..., 12 for the conservation-tillage practices;
n = 1, ..., 28 for the market regions;
p = 1, ..., 4 for the endogenous livestock types;
q = 1, ..., 32 for the livestock rations;
s = 1, 2, 4, ..., 9, 11, ..., 15 for the commodities balanced at the market region;
t = 1, ..., 176 for the transportation activities defined;
u = 1, ..., 15 for the crop.

Livestock production equation:

$$MINL_{np} \leq \sum_p \sum_q Y_{npq} LU_{np} \leq MAXL_{np} \tag{3.9}$$

n = 1, ..., 28 for the market regions;
p = 1, ..., 4 for the endogenous livestock types;
q = 1, ..., 32 for the livestock rations.

National equations. As mentioned previously, the equations that are defined at the national level to balance commodity supply and demand are as follows:

$$\sum_i \sum_j \sum_k \sum_m X_{ijkm} W_{ijkmu} CY_{ijkmsu} \geq CD_s \tag{3.10}$$

i = 1, ..., 105 for the producing areas;
j = 1, ..., 18 for the land groups;
k = 1, ..., 330 for the rotations defined;
m = 1, ..., 12 for the conservation-tillage alternatives;
s = 3, 14 for the commodities cotton and sugar beets;
u = 4, 14 for the crops cotton and sugar beets.

Variables

The variables in the equations are as follows: AD_{ijkm} = the acres of dryland used per unit of rotation k using conservation-tillage method m on land group j in producing area i; AI_{ijkm} = the acres of irrigated land used per unit of rotation k using conservation-tillage method m on land group j in producing area i; AIC_{iu} = the acres of crop u in producing area i as reported in the *1969 Census of Agriculture* (U.S. Department of Commerce, Bureau of Census, 1971); CD_s = the demand for commodity s at the national level; CD_{ns} = the exogenously determined demand for commodity s in market region n; CY_{ijkmsu} = the per acre production of commodity s from crop u in rotation k using conservation-tillage system m on land group j in producing area i; CWU_{iu} = the acre feet per acre water-use coefficient for crop u in producing area i; DA_{ij} = the acres of dryland available on land group j in producing area i; DA_{rs} = the reduction in hay yield associated with the conversion of an acre of irrigated pasture to dryland pasture in water supply region r; DA_{rs} = 0 for all s $\neq$ 5; DC_i = the cost per acre for draining land and converting it to cropland in producing area i; F_n = the number of pounds of nitrogen fertilizer purchased in market region n; FC_n = the cost per pound of nitrogen fertilizer purchased in market region n; HR_i = the proportion of all hay which can be legume hay in producing area i; IA_{ij} = the acres of irrigated land available on land group j in producing area i; IB_r = the acre feet of water transferred out of region r; IC_r = the cost differential on a per acre foot basis for water in region r; LC_{npq} = the cost per unit of livestock activity p receiving ration q in market region n; LD_i = the number of acres of land drained and converted to cropland in producing area i; LDP_{ij} = the proportion of the land drainage in producing area i which is on land group j; LU_{np} = the conversion factor to express the production of livestock-type p consuming ration q in market region n; LW_{npr} = the per unit interaction coefficient for commodity s with livestock-type p consuming ration q in market region n (this will be positive for the livestock products and negative for the ration components); $MINA_{iu}$ = the minimum acreage of crop u required in producing area i; $MINL_{np}$ = the minimum number of units of livestock-type p required in market region n; RC_i = the cost per acre for private irrigation development in producing area i; RD_i = the number of acres developed for irrigation under private development in producing area i; RDP_{ij} = the proportion of the irrigation land developed in producing area i which is in land group j; XC_{ijkm} = the cost per acre of rotation k with conservation tillage practice m in producing area i on land group j; T_{nst} = the number of units of commodity s transported over route t from market region n; TC_{nst} = the cost per unit of commodity s transported over route t from market region n; W_{ijkmu} = the rotation weight for crop u in rotation k using conservation-tillage method m on land group j in producing area i; W_r = the number of acre feet of water purchased in water supply region r; WA_r = the per acre water-release coefficient when converting one acre of irrigated pasture to dryland pasture in water supply region r; WC_r = the cost per acre foot of water purchased in water supply region r; WH_r = the level of irrigated to dryland pasture conversion in water supply region r; WS_r = the per-acre feet of water available for use by the endogenous agricultural sector; X_{ijkm} = the number of acres of rotation k with conservation-tillage m in producing area i on land group j; and Y_{npq} = the level of livestock-type p consuming ration q in market region n.

ORGANIZATION OF THE STUDY

The study was organized around three policy issue areas related to future adequacy of U.S. agricultural resources. To evaluate future resource adequacies using the model developed in this chapter, a base future and several alternative futures are formulated. The Base Alternative represents a continuation of present trends in crop yields, per capita food consumption, exports, etc., for the years 1985 and 2000. In each of the alternative futures one or two parameters in the model are changed with respect to conditions in the Base. The Base and the alternative formulations of the model are then solved for the years 1985 and 2000. Comparison of these solutions provides the basis for the analysis of the adequacy of agricultural resources.

The first policy area considered is the impact of changes in projected export levels on interregional production patterns, land and water use, and prices. Higher export levels are defined in the model while holding all other conditions constant. This High Export Alternative is solved and the results compared to the results from the Base Alternative in the analysis of impacts of changes in export levels.

The second policy area concerns issues of soil and water conservation. For this policy area all the assumptions of the Base Alternative are retained except that (1) the model is constrained to cropping practices that will allow sustained long-run productivity of the nation's agricultural lands, and (2) a higher water-use efficiency is assumed in the model, representing an accelerated adoption rate for water-conserving irrigation technologies. This alternative, referred to as the Resource Conservation Alternative, is solved and the results compared with the results from the solution of the Base Alternative to measure the national and regional impacts of restrictions on soil erosion and a higher water-use efficiency. Also included in this conservation policy area is an Energy Alternative, in which additional water is allocated to energy generation to allow shale oil extraction and to transport and process coal, leaving agriculture with a smaller water supply for irrigation.

The third policy area is concerned with issues related to the enhancement of environmental quality. The environmental variables involved are soil erosion, wetlands development, animal waste disposal, and minimum stream flow to preserve fish and wildlife habitats. Restrictions on these variables are incorporated in the model, then this Environmental Enhancement Alternative is solved and the solution is compared to the Base Alternative to evaluate the impacts on the agricultural sector.

These various alternatives are summarized below in terms of the parameter changes employed to formulate each alternative future with the model. The results and analysis are then presented in the final section of the chapter.

THE ALTERNATIVES

The Base Alternative is the model as described in the previous section outlining the model's major components. The solution to the Base Alternative is utilized as a baseline in the analysis of the other alternatives. The other alternatives are briefly described below.

The High Export Alternative

In this alternative all conditions and constraints of the Base Alternative are retained except that higher levels of exports are used in the model's demand sector. Higher export levels may occur with rising affluence and improved or changing diets in other countries, and a vigorous export policy to stabilize and improve the U.S. balance of trade. A comparison of export levels employed in the model is given in Table 3.5.

The Resource Conservation Alternative

For this alternative, all assumptions of the Base Alternative are retained with two exceptions. The model is constrained to cropping practices to the tolerance (t) values of allowable soil losses as specified by the Soil Conservation Service (SCS) for each producing area and land group. This policy constraint would reflect society's concern for maintaining a productive agricultural land base and also with reducing stream and lake sedimentation. The model also incorporates assumptions concerning an accelerated adoption rate for water-conserving irrigation technologies. The accelerated rate of adoption is reflected by improved water-use efficiency coefficients in the irrigated cropping activities.

The Energy Development Alternative

By the year 2000, water supplies may be withdrawn from agriculture as energy development efforts take priority. Water supplies are therefore adjusted in the model to allow for shale oil extraction and for the transport and processing of coal for generation of electricity. (No adjustment is made in the land base for projected strip mining of coal. With the assumed conversion of strip-mined land back to pasture, the total area under strip mining will not change enough to cause any significant changes in the model solutions.) A comparison of the water supplies available to agriculture for the Base, Energy, and Environmental Enhancement Alternatives is given in Table 3.6.

Table 3.6. Comparison of agricultural water supplies by major river basins for the Base and alternative futures

	Alternative				
	Base		Environmental Enhancement		Energy
River Basins[a]	1985	2000	1985	2000	2000
	(million acre feet)				
Missouri	23.6	23.6	13.4	13.3	23.2
Ark-White-Red	11.7	11.2	5.7	5.2	11.2
Texas-Gulf	10.0	7.7	6.1	3.9	7.7
Rio Grande	4.4	4.4	3.8	3.7	4.4
Upper Colorado	3.0	3.0	.9	.9	2.7
Lower Colorado	5.9	5.5	5.7	5.7	5.4
Great Basin	3.5	3.5	2.7	2.7	3.5
Col-N. Pacific	76.5	76.5	9.1	9.1	76.4
California	30.6	30.6	21.3	21.3	30.5
Western Basins Total[b]	169.4	166.1	68.7	65.8	165.1

[a]See Figure 3.3 for river basin location.

[b]May not add because of rounding.

The Environmental Enhancement Alternative

Several changes are made in the model for this alternative to reflect societal concerns with maintaining water quality and protecting fish and wildlife habitats. In addition to imposing the tolerance levels of allowable soil losses on the model, three additional environmental conditions are included: (1) no further development of wetlands for cultivation is allowed beyond 1975, (2) the water supply available for agricultural uses is reduced sufficiently to allow minimum stream flows for maintenance of water quality and protection of aquatic life (see data in Table 3.6 for available water supplies after adjustments), and (3) livestock wastes must be returned to the land.

RESULTS

Comparison of the model's solutions indicates that the nation's agriculture has a capacity to produce significantly higher levels of output, to significantly reduce soil erosion, and to reduce its impacts on environmental quality. If the increased output or the improved environmental quality are required by the year 2000, the results show that these could be attained with only small increases in consumer food cost. If, however, the achievement of greater output or improved environment is required by 1985, prices will increase sharply and drastic changes will be needed in land use and cropping patterns. The 1985 High Export Alternative comes close to exhausting all available cropland; only 7.2 million acres are not cropped (Table 3.7). Critical assumptions leading to this result for 1985, but not for 2000, include improved agricultural productivity, allowing the model to convert more wetlands and forestland to cropland, to develop more irrigated cropland, and to adjust regional cropping patterns more in 2000 than in 1985. The 1985 Resource Conservation Alternative has 1.7 million more acres of cropland developed than the Base. The improved irrigation coefficients incorporated in the Resource Conservation Alternative, however, reduce irrigation water requirements and result in 2.5 million acre feet of additional surplus water. The 1985 Environmental Enhancement Alternative, for all practical purposes, exhausts not only the supply of land that can be cropped, but also the water supplies available to agriculture. The nearly complete utilization of the agricultural resources occurs because of the joint impact of restricting cropping practices and imposing minimum stream flow requirements to protect wildlife water habitats. The very high supply (or shadow) prices shown in Table 3.8 for the 1985 Environmental Enhancement Alternative and the high cost of food and fiber to consumers (based on the model's supply prices for the commodities) reflect the consequences of limiting U.S. agricultural capacity by such stringent restrictions while holding output requirements constant. The results of the alternatives imply that the simultaneous achievement of the goals to enhance the environment and to expand export levels, as in the Environmental Enhancement and High Export Alternatives, may not be easily attained in 1985.

ble 3.7. Values for major variables at the national level for all alternatives

ternatives	Available Cropland[a]	Cropland Development[b]	Total Available Cropland	Dryland Cropped	Irrigated Acres	Irrigation Development	Land Not Cropped	Water Surplus
	(million acres)							thousand acre-feet
se 1985	369.4	3.8	373.2	286.8	27.2	3.9	59.3	80,817
se 2000	364.4	10.6	375.0	286.5	28.4	6.6	60.1	80,320
gh Export 1985	369.4	7.0	376.4	341.3	27.9	3.9	7.2	79,102
gh Export 2000	364.4	15.8	380.2	323.8	28.9	6.7	27.5	78,473
source Conservation 1985	369.4	5.5	374.9	316.4	28.3	3.9	30.3	83,343
source Conservation 2000	364.4	9.6	374.0	290.0	27.2	6.7	56.8	83,290
ergy 2000	364.4	10.6	375.0	286.3	27.9	6.6	60.8	79,370
vironmental Enhancement 1985	369.4	0.0[c]	369.4	332.3	22.1	3.9	2.7	151
vironmental Enhancement 2000	364.4	0.0[c]	364.4	299.4	22.3	6.7	30.5	1,270

[a]Available cropland after subtracting land required for exogenous crops.

[b]Forest clearing and wetland conversion.

[c]Forest clearing and wetland conversion are not allowed in the Environmental Enhancement alternatives.

Table 3.8. Values for important variables at the national level for all alternatives

Alternatives	Land Shadow Price ($/acre)	Water Shadow Price ($/acre-foot)	Soil Loss (billion tons)	Food Cost Per Captia[a,b] ($)
Base 1985	31.66	11.14	1.3	153.61
Base 2000	23.93	10.14	1.8	124.01
High Export 1985	35.76	23.35	1.3	288.04
High Export 2000	35.09	12.52	1.9	136.58
Resource Conservation 1985	103.33	17.67	0.6	239.41
Resource Conservation 2000	28.65	10.46	0.6	129.15
Energy 2000	24.73	12.59	1.8	124.31
Environmental Enhancement 1985	416.20	222.18	0.5	620.88
Environmental Enhancement 2000	34.58	21.25	0.6	136.49

[a]All prices in 1972 dollars.

[b]Value of commodities based on commodity shadow prices generated by the model.

The following sections provide additional information and analysis for the year 2000 alternatives at a more disaggregated level. For a more detailed presentation of the results for both 1985 and 2000 see Meister et al. (1976).

Adjustments in Production Patterns and Practices

Interregional shifts in crop production. The solution to the Base Alternative provides the starting point for the analysis of the interregional shifts found with the various alternatives. A regional breakdown of U.S. totals for feed grains, wheat, and soybean acreage is provided in Table 3.9. Most of the feed grains in the Base Alternative are grown in the Northeast and Mid-Continent regions. Most of the wheat is grown in the Mid-Continent region. Large acreages of soybeans are produced in all three of these regions.

In the High Export Alternative, the Mid-Continent region shows large increases in crop acreages as a high proportion of the extra production needed to meet the higher export levels comes from this area. This result implies that the Mid-Continent region has considerable uncropped land and unused water resources in the Base that are relatively productive compared to other areas of the United States. In the Resource Conservation Alternative, the total U.S. acreage of soybeans declines significantly compared to the Base. This decline is the result of a decreased use of soybean meal as a protein source in the livestock rations. The soybean meal in the rations is reduced because of its high price relative to other protein sources in the Resource Conservation Alternative. The higher soybean price is the result of the relatively high cost of protecting the land from soil

Table 3.9. Regional crop acreage by alternative for the year 2000

Reporting Zones and Crops	Base Acres	Base Index of Base	High Export Acres	High Export Index of Base	Resource Conservation Acres	Resource Conservation Index of Base	Environmental Enhancement Acres	Environmental Enhancement Index of Base
	(thousand)		(thousand)		(thousand)		(thousand)	
Northeast[a]								
Feedgrains	50,283	100	56,238	112	53,097	106	48,586	97
Wheat	6,827	100	7,677	112	7,056	103	8,256	121
Soybeans	42,144	100	43,598	127	37,142	88	39,200	93
Southeast[b]								
Feedgrains	2,763	100	3,341	121	7,519	272	3,947	143
Wheat	3,718	100	4,620	124	1,415	38	1,839	49
Soybeans	26,378	100	27,097	103	13,814	52	17,316	66
Mid-Continent[c]								
Feedgrains	33,845	100	34,327	101	22,774	67	40,643	120
Wheat	26,655	100	39,352	148	64,038	240	25,613	96
Soybeans	18,694	100	23,994	128	11,232	60	28,277	151
West[d]								
Feedgrains	2,932	100	3,001	102	4,444	152	2,972	101
Wheat	4,844	100	7,279	150	8,487	175	6,945	143
Soybeans	0	100	0	100	0	100	0	100
United States								
Feedgrains	89,828	100	97,007	108	113,143	126	96,157	107
Wheat	41,962	100	58,936	140	58,231	139	42,662	102
Soybeans	87,219	100	94,694	109	61,194	70	85,701	98

[a]Includes river basins of New England, Mid Atlantic, Great Lakes, Ohio, and Upper Mississippi.

[b]Includes river basins of South Atlantic-Gulf, Tennessee, and Lower Mississippi.

[c]Includes river basins of Souris-Red-Rainy, Missouri, Arkansas-White-Red, and Texas-Gulf.

[d]Incluces river basins of Rio Grande, Upper Colorado, Lower Colorado, Great Basin, Columbia-North Pacific, and California.

erosion when soybeans are grown. Soybean acreage declines uniformly across the major soybean producing areas.

There are significant interregional shifts associated with feed grains and wheat when the Resource Conservation Alternative is compared to the Base. In particular, the Mid-Continent region shows a large change in the mix of crops grown. There is a dramatic shift of wheat production, away from both feed grains and soybeans. Compared to the Base, national totals for the acreages of the crops in the Environmental Enhancement Alternative do not change very much. Feed grain acreage does increase some, mostly in the Mid-Continent region. There is a significant interregional shift of soybean production from the Southeast region to the Mid-Continent region.

Wetland and forestland conversion. The model is formulated to augment the already available cropland through the conversion of wetlands and forestlands into cropland. This conversion occurs only if such developed cropland has a comparative advantage in producing agricultural crops within the interregional competition framework of the model, i.e., it is cheaper to produce on the converted land (with conversion costs included in cost of production) than to use already available idle cropland elsewhere in the country. As shown in Table 3.7, several million acres of land are converted to cropland under all alternatives except the Environmental Enhancement Alternative in which conversion is not allowed as a policy measure to protect wildlife habitats. More than 15 million acres are converted in the High Export 2000 Alternative, a 50 percent increase over the acreage converted in the Base 2000 Alternative.

When these aggregate figures are considered at the river-basin level, it is clear that the conversion is not spread evenly across the country (Table 3.10). The High Export Alternative indicates that most of the additional converted land to meet the increased export requirements is developed in the South Atlantic-Gulf river basin. (See Figure 3.3.) In the High Export Alternative, land conversion in every region is equal to or greater than that in the Base. This is not the case under the Resource Conservation Alternative, where the Missouri river basin has the largest increase in converted acres; while in the Lower Mississippi and the Arkansas-White-Red river basins, significantly fewer acres are converted than in the Base. The imposition of the soil and water conservation measures on the model results in very different effects across the river basins. The Energy Alternative produces the same land conversion result as the Base.

Table 3.10. Acres of land developed by major river basins for the Base and other alternatives for the year 2000[a,b]

River Basins	Alternative: Base	High Export	Resource Conservation	Energy
	(thousand acres)			
New England	0	987	0	0
Mid Atlantic	77	77	77	77
S. Atlantic Gulf	2,808	5,459	2,013	2,808
Great Lakes	1,113	1,113	1,113	1,113
Ohio	614	1,083	800	614
Tennessee	232	232	232	232
Upper Mississippi	1,315	1,673	1,349	1,315
Lower Mississippi	2,258	2,258	1,693	2,258
Souris-Red-Rainy	462	462	462	462
Missouri	607	955	1,248	1,112
Ark.-White-Red	1,112	1,248	0	0
Texas-Gulf	0	288	0	0
Rio Grande	0	0	0	0
Upper Colorado	0	0	0	0
Lower Colorado	0	0	0	0
Great Basin	0	0	0	0
Col.-N. Pacific	0	0	0	0
California	0	0	0	0
United States[c]	10,602	15,841	9,369	10,602

[a]Land developed is the total acres of wetlands drained and forestlands cleared.

[b]No development is allowed for the Environmental Enhancement Alternative.

[c]May not add due to rounding.

Soil erosion and farming practices. Estimated soil loss for the major U.S. river basins under each alternative are presented in Table 3.11. For the nation and most river basins, soil loss increases only slightly under the High Export Alternative compared with the Base. Total erosion losses actually decrease in a few river basins because of the interregional redistribution of crops and because of fewer acres of soybeans and silages. Nationally, agriculture apparently can respond to the larger export demands without significant additional soil erosion.

Table 3.11. Levels of soil loss by major river basin for the year 2000

	Alternative			
River Basins	Base	High Export	Resource Conservation	Environmental Enhancement
	(million tons)			
New England	0.8	6.3	0.3	1.4
Mid Atlantic	46.8	41.2	14.6	14.3
South Atlantic-Gulf	223.0	233.7	47.6	44.1
Great Lakes	45.6	44.1	32.6	30.4
Ohio	112.3	133.0	57.4	57.1
Tennessee	20.4	23.2	5.9	5.1
Upper Mississippi	447.4	434.7	150.8	144.2
Lower Mississippi	282.9	300.4	37.9	33.6
Souris-Red-Rainy	34.3	36.2	19.2	18.5
Missouri	315.2	334.5	133.7	130.1
Arkansas-White-Red	117.7	140.6	57.2	62.7
Texas-Gulf	88.0	105.2	29.0	37.5
Rio Grande	4.7	6.9	3.4	3.9
Upper Colorado	1.9	2.0	1.5	1.3
Lower Colorado	0.7	0.7	0.7	0.5
Great Basin	4.0	4.4	2.4	2.2
Columbia-North Pacific	47.9	38.5	20.4	18.7
California	3.9	4.6	3.9	3.5
United States[a]	1,797.6	1,890.3	618.5	609.1

[a]May not add because of rounding.

The Base and the High Export Alternatives produce soil loss levels that are comparable to present actual losses. With erosion reduced to tolerance values in the Resource Conservation Alternative, national soil loss declines by about two-thirds. The reduction in soil loss is not uniform across all regions because of varying soil and climatic characteristics. Those regions in the United States with abundant rainfall, warm climate, and soils susceptible to erosion are required to make more adjustments in cropping practices and erosion-control measures to attain the soil-loss tolerance values. Thus total soil losses in such regions decline more than for the nation as a whole. The Environmental Enhancement Alternative produces only slight additional reductions in soil loss compared with the Resource Conservation Alternative. The reductions that do occur stem from changes in water use and irrigation.

The impact of the soil-loss restraint in the Resource Conservation Alternative is clearly shown in the large reduction in the total amount of soil eroded. (See Table 3.11.) Even more noticeable are the large reductions in the South Atlantic-Gulf, Upper and Lower Mississippi, and Missouri basins. Achieving these large reductions in soil loss requires changes not only in regional land use and crop patterns, but also in conservation and tillage practices. Straight-row cropping decreases substantially while increases occur in contouring and terracing. Also, minimum tillage practices increase (Table 3.12).

Table 3.12. Acreages under the various conservation and tillage practices for the year 2000

Item	Base	Alternative: Resource Conservation	Alternative: Environmental Enhancement
	--------(percent of harvested acres)------		
Tillage Practice:			
Conventional tillage, residue removed	18.0	17.0	19.0
Conventional tillage, residue left	48.4	42.6	41.4
Minimum tillage	33.5	40.4	41.6
Conservation Practice:			
Straight-row	39.8	23.8	20.2
Contouring	31.6	36.6	37.2
Strip cropping	14.3	14.5	13.9
Terracing	14.3	25.6	28.8

Livestock production. In the Base, U.S. hog production is concentrated in the Northeast region (Table 3.13). Beef feeding is spread more uniformly across the regions than hog production. The model produces significant interregional shifts of both hog production and cattle feeding under the various alternatives. Hog production shifts slightly, compared to the Base, from the Northeast region to the Mid-Continent region under both the High Export and the Environmental

Table 3.13. Regional distribution of beef feeding and hog production for the year 2000

Zone	Base: Beef feeding	Base: Hogs	High Export: Beef feeding	High Export: Hogs	Resource Conservation: Beef feeding	Resource Conservation: Hogs	Environmental Enhancement: Beef feeding	Environmental Enhancement: Hogs
	------------------------(percent of U.S. total)------------------------							
Northeast	13.2	83.0	14.8	80.1	15.7	83.0	14.5	77.1
South	20.9	2.4	26.0	2.4	9.7	2.4	7.6	2.4
Mid-Continent	49.4	14.3	42.4	17.3	59.3	14.3	64.3	20.3
West	16.6	.2	16.8	.3	17.1	.3	13.7	.3
United States[a]	100.0	100.0	100.0	100.0	100.0	100.0	100.0	100.0

[a]May not add due to rounding.

Enhancement Alternatives. Cattle feeding shifts dramatically to the Mid-Continent region from the South under the Resource Conservation and the Environmental Enhancement Alternatives.

In addition to interregional shifts of production, the model allows for changes in the livestock ration as higher-cost feedstuffs are substituted with lower-cost feedstuffs (Table 3.14). For example, the use of silage as a feedstuff for cattle feeding increases in the High Export Alternative as the productive capacity of American agriculture is more fully utilized. In the Resource Conservation and Environmental Enhancement Alternatives, however, the use of silages in the livestock rations declines compared to the Base. This decrease is due to the very erosive characteristics of silage crops and thus the high cost of protecting the land from soil erosion as required under both of those alternatives.

le 3.14. Average livestock rations for the Base and other alternatives for the year 2000

estock s and ernative	Feedstuffs								
	Corn	Sorghum	Barley	Oats	Wheat	Oil-meals	Legume hay	Nonlegume hay	Silage
	---------(bushels)---------					(cwt.)	---------(tons)---------		
f feeding									
e	6.92	1.97	3.81	0.00	0.00	2.19	0.10	0.17	5.42
a Export	2.53	2.37	0.28	0.00	0.00	2.36	0.03	0.07	6.51
urce onservation	9.57	5.13	5.50	0.00	0.00	1.94	0.18	0.20	4.46
ronmental hancement	7.40	2.87	5.75	0.00	0.00	2.08	0.13	0.23	4.84
ry									
e	112.77	2.83	4.84	6.96	0.56	6.38	1.49	1.56	1.66
a Export	115.23	0.00	3.74	5.84	0.40	6.57	1.44	1.67	1.62
urce nservation	115.60	0.00	4.46	6.36	0.48	6.31	1.53	1.53	1.71
ronmental hancement	110.50	0.17	9.27	7.50	0.74	6.51	1.38	1.77	1.51
:									
e	5.74	0.37	0.003	0.03	0.06	0.50	0.006	0.00	0.00
a Export	5.84	0.33	0.003	0.03	0.002	0.50	0.006	0.00	0.00
urce onservation	5.78	0.36	0.003	0.03	0.06	0.51	0.007	0.00	0.00
ronmental hancement	5.66	0.46	0.003	0.03	0.06	0.50	0.006	0.00	0.00

Adequacy of Water Supplies

The evaluation of the adequacy of water supplies for agriculture in this study is based on two concepts. Water supplies are adequate if projected food and fiber demands in the model are met and if the water supply in each producing area allows land presently irrigated to be irrigated in the future. This latter concept is incorporated into the model with the restraint that the total irrigated acreage in each producing area be at least equal to the actual 1969 level for endogenous crops. There are three producing areas where projected depletion of groundwater supplies makes it impossible to entirely irrigate 1969 acreages in the year 2000. All three producing areas are in the Texas High Plains. (Consequently, the irrigated-acres restraint in the model in these producing areas is reduced to a level consistent with projected water supplies in the year 2000 to make the model feasible.)

The Base and the four alternatives determined that both total water supplies and regional water supplies are adequate to meet the projected level of food and fiber demand levels for the year 2000 in all alternatives except in the Environmental Enhancement Alternative. The Environmental Enhancement Alternative indicates that the simultaneous achievement of the objectives of enhancing the environment and meeting the projected demand levels for the year 2000 is not possible because of water deficits in some areas in the West. This is an important result, but one that has to be viewed in light of the assumptions for this alternative as well as the various restraints imposed on the model concerning interregional adjustments in cropping patterns. The crucial assumptions are the high priority given to water demands for fish and wildlife and the land-use restraints that the irrigated acreages must be at least equal to the actual 1969 irrigated acreage. Some of the minimum stream flow requirements are several times larger than the projected deficits within some producing areas in the Columbia-North Pacific and the Texas-Gulf river basins. (See Figure 3.3.) Hence, small reductions in minimum stream flow requirements will allow simultaneous achievement of a slightly lower level of environment enhancement while meeting projected demands.

Total water use for irrigation in each river basin is shown in Table 3.15. The Base, High Export, and Resource Conservation Alternatives all have identical water supplies for agriculture. The Energy Alternative and the Environmental Enhancement Alternative have reduced water supplies. The Energy Alternative's water supply is reduced because water for energy development is given a higher priority than agriculture. Therefore this water requirement is subtracted from the model's available water supply for agriculture. In the Environmental Enhancement Alternative, fish and wildlife are given the highest priority and requirements to maintain minimum stream flow levels reduce the amount available for agriculture.

Table 3.15. Water used for irrigation in all alternatives for the year 2000

River Basins	Alternative				
	Base	High Export	Resource Conservation	Energy	Environmen Enhanceme
	(thousand acre-foot)				
Missouri	16,656	16,847	16,599	16,618	12,614
Ark.-White-Red	6,165	6,440	5,849	6,161	4,196
Texas-Gulf	5,865	5,937	5,248	5,865	4,561
Rio Grande	3,943	4,578	3,821	4,252	638
Upper Colorado	3,008	3,009	3,008	2,718	895
Lower Colorado	4,967	5,013	4,725	4,930	5,136
Great Basin	3,454	3,449	3,449	3,458	2,630
Columbia-N. Pacific	17,179	17,197	16,290	17,166	8,985
California	22,653	23,506	21,911	22,646	20,010
Western	83,890	85,688	80,910	83,826	62,677

The water supply in two river basins in particular, the Columbia-North Pacific and the Texas-Gulf, is substantially reduced. The large reductions in water available for agriculture in these regions is responsible for the deficits mentioned earlier.

Total agricultural water use in the Base is 85.7 million acre feet. In the High Export Alternative, agricultural water use increases to 87.6 million acre feet as the irrigated crop acreage is significantly increased. At the river-basin level in the High Export Alternative all but three regions show water surpluses. The three exceptions are the Texas-Gulf, the Upper Colorado, and the Great Basin river basins. The latter two basins exhaust all their water under the conditions of the Base. The increase in agricultural water use in the Texas-Gulf river basin is due to the increase in the irrigated sorghum grain acreage. Total agricultural water use under the Resource Conservation Alternative decreases from the Base by 3.4 percent to 82.7 million acre feet. This decrease is mainly the result of higher water-use efficiency assumed for this alternative. The regional water-use pattern for the Resource Conservation Alternative shows larger surpluses in all regions when compared with the Base. There is even a small surplus in the Great Basin. This alternative shows clearly the effect of the higher water-use efficiency. The achievement of this higher efficiency would substantially help to reduce concerns about adequacies of future water supplies for agriculture. The Energy Alternative has a slightly different mix of irrigated crops. Irrigated corn acreage, for example, is reduced substantially. This different crop mix results in different regional water-use patterns. Part of this reallocation is the result of the reduced water supply in the Upper Colorado, where water was already in short supply in the Base. Still, the overall effect is minimal.

The changes in the supply and demand for water following the changes in the parameters of the model are reflected in the prices for water (and land) resources and for the commodities produced in the alternatives. The prices for water are greatly influenced by availability of water and by the requirements for agricultural commodities incorporated in the model (Table 3.16). In this study, the Environmental Enhancement Alternative produced the largest increase in average

e 3.16. Shadow prices for water by major river basin for the Base and the other alternatives for the year 2000[a]

	Alternative				
r Basins	Base	High Export	Resource Conservation	Energy	Environmental Enhancement
	(dollars per acre-foot)				
ouri	16.41	19.62	12.58	16.62	26.27
nsas-White-Red	34.98	39.71	23.08	34.88	44.47
s-Gulf	17.80	22.01	20.06	20.85	29.18
Grande	8.70	11.06	6.94	8.68	22.82
r Colorado	6.63	7.40	6.45	6.61	17.39
r Colorado	7.73	9.39	7.71	7.74	8.02
t Basin	8.48	9.33	8.98	8.47	17.19
mbia-N. Pacific	2.89	5.06	2.54	2.88	6.92
fornia	10.65	10.65	10.61	10.65	21.22
ern Basins	12.52	14.31	10.46	12.59	21.25

[a]Values are in 1967 dollars.

shadow prices for water. These price increases are not uniform across the river basins, as the price of water changes only slightly in the Lower Colorado river basin, but more than doubles in the Rio Grande river basin. The prices for water under the Energy Alternative are about equal to the Base, while the price for water increases only slightly under the High Export Alternative. The average price of water declines, however, under the Resource Conservation Alternative relative to the Base because of the assumed improvement in water-use efficiency.

Regional Income Distribution

Changes in returns to resources, indicated by comparison of the results from the various alternatives, imply income and wealth changes for individuals in agriculture. Indices of regional changes in returns to land, labor, and water from the study are presented in Table 3.17 (with the Base equal to 100). The results range from the small changes in the Energy Alternative to the relatively large changes in the High Export and the Environmental Enhancement Alternatives. For instance, the Lower Mississippi river basin gains under the High Export Alternative but loses under the Resource Conservation and Environmental Enhancement Alternatives. In the New England basin, returns increase dramatically under both the High Export and Environmental Enhancement Alternatives. This increase is the result of the large increase in crop acreage under these two alternatives. The Rio Grande basin is sensitive to available water supplies, and in both the Energy and Environmental Enhancement Alternatives returns to agriculture resources

Table 3.17. Comparison of the relative changes in the total of the returns to land, water, and labor resources in crop production by major river basin for all alternatives for the year 2000[a]

	Alternative			
River Basins	High Export	Resource Conservation	Energy	Environmental Enhancement
	(Base Alternative=100)			
New England	793	100	101	293
Mid Atlantic	156	109	100	141
South Atlantic-Gulf	155	91	99	115
Great Lakes	145	116	99	137
Ohio	192	154	128	187
Tennessee	139	126	99	155
Upper Mississippi	142	114	100	138
Lower Mississippi	133	91	99	95
Souris-Red-Rainy	191	145	102	195
Missouri	154	122	103	136
Arkansas-White-Red	149	97	102	127
Texas-Gulf	131	104	98	147
Rio Grande	136	107	117	208
Upper Colorado	117	101	96	36
Lower Colorado	114	107	98	128
Great Basin	132	106	95	136
Columbia North Pacific	148	104	95	98
California	127	112	103	99
United States	150	113	103	156

[a]Returns are based on shadow prices generated by the model.

decline. There are many more of these changes that are important. Each alternative has beneficial and adverse effects on the individual regions. In evaluating the alternative futures these effects should be taken into account.

SUMMARY

The economic and related impacts of higher exports, erosion control, and environmental enhancement alternatives for agriculture indicate that higher than projected exports can be attained with slight increases in soil loss. In addition, a program to reduce erosion losses to levels that will allow soil productivity to be maintained would also allow projected export levels to be attained. Environmental enhancement in the form of simultaneous reductions in soil loss and improved stream flows would, however, compete with higher exports.

While the environmental enhancement program analyzed would have the greatest impact on average water values, high exports and resource conservation programs would similarly increase average returns to water resources. Important differences prevail in the interregional distribution of income and resource values, however. Higher exports result in reasonably uniform increases in resource values in all producing areas. A land and water conservation program, however, would tend to redistribute income in favor of areas with modest rainfall and level land. An environmental enhancement program would tend to bring greater income and resource values to the dryland farming regions not subject to erosion at the expense of some irrigated areas using surface water and of regions with high rainfall and erosive lands.

REFERENCES

Meister, A. D., and K. J. Nicol. 1975. A Documentation of the National Water Assessment Model of Regional Agricultural Productions, Land and Water Use and Environmental Interaction. Center for Agricultural and Rural Development, Iowa State University, Ames.

Meister, A. D., E. O. Heady, K. J. Nicol, and R. W. Strobehn. 1976. U.S. Agricultural Production in Relation to Alternative Water, Environmental, and Export Policies. CARD Report 65. Center for Agricultural and Rural Development, Iowa State University, Ames.

National Water Commission. 1973. Water Policies for the Future. U.S. Government Printing Office, Washington, D.C.

U.S. Department of Agriculture, Conservation Needs Inventory Committee. 1971. National Inventory of Soil and Water Conservation Needs, 1967. USDA Statistical Bulletin 461. Washington, D.C.

U.S. Department of Commerce, Bureau of Census. 1971. *U.S. Census of Agriculture, 1969.* Vol. 1: *Statistics for the States and Counties.* U.S. Government Printing Office, Washington, D.C.

U.S. Water Resources Council. 1974a. 1972 OBERS Projections, Vols. 1-7. U.S.

Government Printing Office, Washington, D.C.
___. 1974b. 1975 Water Assessment-Nationwide Analysis Work Statement. U.S. Government Printing Office, Washington, D.C.
___. 1975. 1975 OBERS Projections, Agricultural Supplement. U.S. Government Printing Office, Washington, D.C.
Weinberger, M. L., and H. L. Hill. 1971. Benefit-Cost Relationships in Sediment Control. Presented at the 26th Annual Meetings of the Soil Conservation Society of America, New York, Mimeograph.
Wischmeier, W. H., and D. D. Smith. 1965. Predicting Rainfall-Erosion Losses from Cropland East of the Rocky Mountains. USDA Agriculture Handbook 282. Washington, D.C.

CHAPTER 4. ANALYSIS OF SOME ENVIRONMENTAL POLICIES FOR AMERICAN AGRICULTURE

by Gary F. Vocke and Earl O. Heady

THIS STUDY is another in the sequence by the Center for Agricultural and Rural Development (CARD), under a grant from the National Science Foundation's Research Applied to National Needs (RANN), concerned with policies for resource use in agriculture. The objective of this study is the analysis of a set of policies designed to alter current agricultural practices so as to reduce U.S. agriculture's impact on the environment. The issues addressed by the policies include reducing soil erosion in the United States to the soil-loss tolerance levels discussed in previous chapters. These rates range from 1 to 5 tons per acre per year, depending upon soil properties, soil depth, topography, and previous erosion.

A second environmental policy considered is a policy restricting the use of nitrogen fertilizer. The role of nitrogen in agriculture has attracted special attention because of its alleged potential to harm the environment. Farmers have greatly increased the use of nitrogen fertilizer because of the high crop yields that can be obtained at a relatively low cost. This expanded use is viewed by some persons as a threat to the quality of surface and underground water supplies. Nitrogen in the nitrate form is easily leached from the soil into water supplies, where excessive amounts may stimulate the growth of algae and can be a health hazard. Nitrogen can also enter water supplies via eroded sediments. Factors affecting the level of soil erosion will also affect the level of nitrogen, as well as other nutrients, entering the surface water supplies of the nation.

A third environmental policy investigated in this study concerns the restriction on the use of certain organo-chlorine insecticides. Insecticides are widely used because of their effectiveness in controlling insect infestations. With some organo-chlorine insecticides, this widespread use becomes a matter of concern because their persistence allows the residues to build up in the soil and to cycle through the environment, harming other forms of life.

The final policy included in the set of environmental policies analyzed here concerns regulations requiring the control of runoff from livestock feedlots. When livestock are confined in feedlots and their wastes allowed to accumulate, a potential for pollution of waterways exists. Rainfall or snowmelt produces feedlot runoff, which can carry nutrients and organic matter from the lots into streams. The organic matter creates a water quality problem as its decomposition reduces the amount of oxygen in the water. Fish and other water life are then affected. The nutrients carried into the waterways may stimulate excessive growth of algae and weeds.

This set of environmental policies are evaluated using an interregional linear programming model based on the model developed for the National Water Assessment (NWA), reviewed in Chapter 3. This chapter provides a brief overview of this model and then presents a summary of the results of the analyses of the policies considered.

THE MODEL

The construction and use of the linear programming model on which the study is based is summarized here in four parts: availability of land and water resources, crop and livestock production activities for the transformation of these resources into agricultural commodities, a commodity transportation network, and domestic and foreign demands for agricultural products. This model determines the organization of U.S. agriculture that will minimize the cost of meeting the domestic and foreign demands for agricultural products under each environmental policy. It assumes all resources used in agriculture, except land and water, receive their market rate of return. The returns to land and water are determined endogenously in the model.

Regions of the Model

This study uses the same 105 producing areas, 51 water supply regions, and 28 market regions used for the model in Chapter 3. These regions are shown in Figures 3.1 and 3.2. However, a different set of reporting regions than those in Figure 3.3 are used in this chapter. We use the 12 reporting regions shown in Figure 4.1. The reporting regions shown in Figure 4.1 are formed by aggregating contiguous market regions. These reporting regions are used for presenting the results of the analysis.

Land Base

Models of previous chapters used 9 land groups per producing area. However, since not all of these details could always be utilized, and in order to reduce computation costs, the number of land groups per producing area is reduced to 5 in this study. The construction of the 5 land groups from the twenty-nine land classes and subclasses explained in Chapter 1 are summarized in Table 4.1. The county acreages are aggregated, for dryland and irrigated uses, to the 105 producing regions by the twenty-nine capability class-subclasses. These twenty-nine class-subclasses are then aggregated into 5 land quality groups to create the land base for the model.

Crop Production Sector

The endogenous crop production sector includes alternative production activities for grain sorghum, sorghum silage, barley, corn, corn silage, cotton,

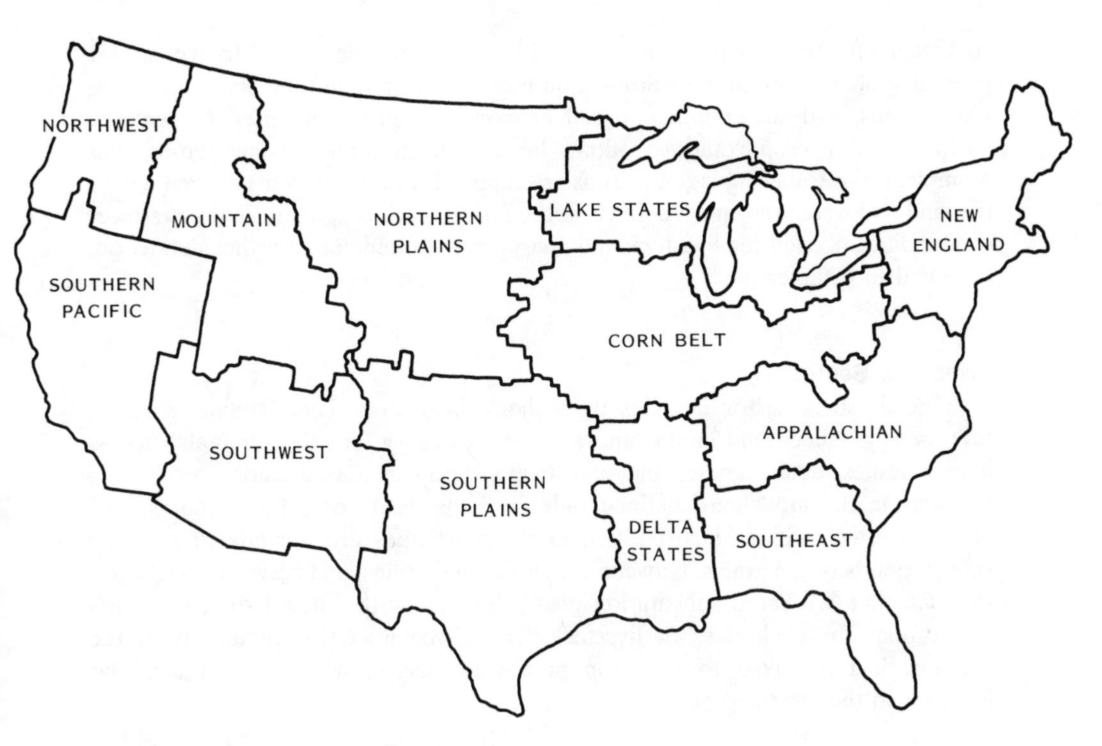

Figure 4.1. The 12 reporting regions.

Table 4.1. Land quality class and subclass aggregations to the five land quality groups for the model's land base

Land Quality Group	Inventory Class-Subclass	Acres
1	I, IIwa,[a] IIIwa	64,596,000
2	rest of II, III, IV, all of V	213,385,000
3	IIIe	71,001,000
4	IVe	29,886,000
5	VI, VII, VIII	14,340,000

[a]wa means that the drainage problem has been treated.

legume and nonlegume hay, oats, soybeans, sugar beets, and wheat consistent with the production possibilities of each producing area. (The remainder of the crops grown in the U.S. are exogenous, as explained in earlier chapters.) Cropping activities for the endogenous crops are defined for each land group in each producing area specifying alternative rotations, tillage and conservation practices, and irrigated or dryland farming. Each rotation is combined with one of four conservation practices: straight-row cropping, contouring, strip cropping, or terracing. Conservation practices are defined on land quality groups according to recommendations provided by the Soil Conservation Service (SCS). A cropping activity is completed by adding one of three tillage practices: conventional tillage with residue removed, conventional tillage with residue left, or reduced tillage. The coefficients for these activities are created by procedures developed to account for producing-area differences in production cost, fertilizer requirements, crop yields, water needs, and susceptibility to soil erosion. The procedure used to generate coefficients for crop rotations allows for interrelationships among crops. For example, crops following legume crops are supplied with carry-over nitrogen. Using the land and water resources defined in the model, each cropping activity produces commodities needed for livestock and consumer requirements. Further details can be found in Chapter 3.

Livestock Sector

The livestock sector includes dairy, hogs, beef cows, beef feeding, broilers, turkeys, eggs, sheep and lambs, and a general category for other animals such as horses, mules, ducks, geese, and zoo animals. Endogenous livestock production activities in the model are defined only for hogs, beef cows, beef feeding, and dairy. Livestock rations incorporated in these activities are formulated to allow substitution between grains, between roughages and grains, and between roughages. (See Chapter 3.) These substitution possibilities allow the model to select least-cost rations for each class of livestock. The nitrogen in the manure from the livestock is transferred to the crop production sector, where it is utilized as fertilizer in the cropping activities.

Water Sector

The water sector of the model defines the availability of water to agriculture in the western United States in producing areas 48-105 and is the same as that illustrated in Figure 3.1. Additional information about the water supplies and the transfer activities can be found in Meister and Nicol (1975).

Transportation Sector

Transportation routes are defined between all contiguous market regions shown in Figure 3.2. The length of each route is the distance between the metropolitan centers in each market region. Some heavily used long-haul routes between noncontiguous market regions are defined if these routes reduce mileage by 10 percent over accumulated short-travel routes. For each route two activities are defined for each commodity, one for shipment in each direction.

Commodity Requirements Sector

The commodity requirements sector specifies the production of the agricultural commodities to meet projected levels of demand for food and fiber, net exports, exogenous livestock feed requirements, and the industrial and nonfood uses. These requirements are based on 1985 projections and are defined at the market-region level. These projections are given in Tables 3.4 and 3.5.

Time Horizon

Evaluation of policy impact alternatives within the limitations of the model requires that a sufficient time horizon be specified to allow for the implied adjustments to materialize. In this study, 1985 was selected as the year of analysis. Activities defined in the model are consistent with projected and expected production options for 1985. Resource availabilities and commodity requirements in the model are also consistent with projections for 1985.

ALTERNATIVE FUTURES

Seven alternative formulations of the model for the year 1985 are considered here to analyze the effects that conservation and environmental improvement policies might have on U.S. agriculture. The alternatives are: (1) Base Alternative, where ongoing trends are assumed and no environmental restraints are imposed; (2) Soil Conservation Alternative, where ongoing trends are the same as in the Base Alternative but soil erosion is restricted; (3) Nitrogen Restriction Alternative, where ongoing trends are the same as in the Base Alternative but no more than 50 pounds of nitrogen can be applied per acre on any crop; (4) Insecticide Restriction Alternative, where ongoing trends are the same as in the Base Alternative but farmers are denied the use of the organo-chlorine insecticides Chlordane and Heptachlor; (5) Feedlot Runoff Control Alternative, where ongoing trends are the same as in the Base Alternative but feedlot operators are required to control the runoff from their feedlots; (6) High Export Alternative, where all available cropland is planted and no environmental restraints are imposed; and (7) Restricted Export Alternative, where the soil loss, nitrogen, insecticide, and feedlot runoff restrictions of the previous alternatives are all simultaneously imposed on the High Export Alternative.

METHODOLOGY

The analysis of the alternatives is carried out in the following manner: First, the Base Alternative, which does not include any environmental restrictions on agriculture, is solved under the demand projections cited above. Next, the alternatives, which include the selected environmental restraints, are solved under these demand levels. Then, the solutions from the Base Alternative and the environmental alternatives are compared and any differences are attributed to the policy restraint imposed on the model.

ANALYSIS OF THE SOIL CONSERVATION ALTERNATIVE

The Base Alternative is formulated with no restrictions on the selection of cropping practices, regardless of the effect on soil erosion. The Soil Conservation Alternative is formulated so that soil erosion rates will be less than soil-loss tolerance rates, the t values set by soil scientists. Soil scientists estimate that the amount of soil that can be lost by erosion without impairing future productivity varies from 1 ton per acre per year on shallow soils to 5 tons on deep soils. Modifying the model to reflect a policy requiring agriculture to limit soil erosion to these levels in 1985 eliminates those cropping activities that do not provide adequate protection for the soil. As described earlier, the cropping activities in the model are defined by land quality group in each producing area and specify a combination of rotation, tillage, and conservation practices. The rate of soil erosion for each of these cropping activities is determined with the Universal Soil-Loss Equation (Wischmeier and Smith, 1965). To develop the Soil Conservation Alternative, each cropping activity in the model is checked, and all activities with erosion rates greater than the soil-loss tolerance levels are removed from the model.

The results from the two alternatives indicate that agriculture has the potential to adjust to meet the projected domestic and foreign demands while restricting soil loss to t values. The adjustments for the agricultural sector implied in the comparison of these two alternatives include changes in rotations, tillage, and conservation practices employed to produce the crops and the regional pattern of crop production. These changes, in turn, alter the commodity prices and the returns to agricultural resources in the model. These adjustments in cropping practices and patterns and changes in commodity prices and resource returns are briefly discussed in the following sections.

Changes in the Crop Production Practices

The crop production practices in the Soil Conservation Alternative are significantly different from those in the Base Alternative. The proportion of acres protected with reduced tillage increases under the Soil Conservation Alternative (Table 4.2). There is less straight-row and contour farming and more strip cropping and terracing (Table 4.3). Finally, the percent of the rotation sequences

Table 4.2. Comparison of the percentage of acres by tillage practice in the Base and the Soil Conservation Alternatives

Tillage Practice	Base	Soil Conservation Alternative
Conventional tillage with:		
Residue removed	15	12
Residue left	25	21
Reduced Tillage	60	67

Table 4.3. Comparison of percentage of land by conservation practice in the Base and the Soil Conservation Alternatives

Conservation practices	Base	Soil Conservation Alternative
Straight row and contour farming	92	78
Strip cropping and terracing	8	22

that are row crops declines (Table 4.4). Crops such as corn and soybeans that provide relatively inadequate protection for the soil are increasingly grown in rotations with small grain and hay crops to provide more protection for the soil in the Soil Conservation Alternative.

These adjustments in the cropping patterns occur because of substitutions between feedstuffs in the livestock rations. When the row crops, e.g., silage crops and soybeans, are grown using conservation practices such as strip cropping and terracing, the cost of production increases relative to small grains and hay. Consequently, there is a 31 percent decline in the acreage of silage and increases of 34 and 33 percent, respectively, in the acreage of small grains and hay in the Soil Conservation Alternative compared to the cropping patterns of the Base Alternative.

Table 4.4. Comparison of percentage distribution of row crop acres by rotation sequence in the Base and the Soil Conservation Alternatives

	Percentage of Rotation Sequence that is Row Cropping[a]			
Alternatives	25	50	75	100
	(percent)			
Base	18	29	11	41
Soil Conservation	33	27	15	22

[a]The numbers 25, 50, 75, and 100 represent rotations which have these percentages of the land devoted to row crops. Thus, under the Base Alternative, 18 percent of the land is in a rotation which has 25 percent of the land in row crops.

These changes in the crop mix substantially alter the use of agricultural resources from the Base Alternative (Tables 4.5 and 4.6). The use of land, nitrogen fertilizer, and pesticides allocated to the production of the small grains and the hays increases. The use of these inputs in the production of the silages and soybeans declines.

Table 4.5. Percentage change of output and inputs by commodity group in the Soil Conservation Alternative compared to the Base

	Changes from the Base[a]			
Commodity groups	Production	Acres	Nitrogen	Pesticide expenditures
	(percent)			
Corn and Sorghum Grain	-3	4	5	2
Barley, Oats, and Wheat	32	34	48	67
Cotton	0	-3	22	0
Soybeans	-6	-4	113	8
Legume and Nonlegume Hay	15	33	48	262
Corn and Sorghum Silage	-35	-31	-28	-30

[a]A positive (negative) value indicates a percentage increase (decrease) in the Soil Conservation Alternative compared to the Base. A breakdown of resource use in the Base is provided in Table 4.6.

Table 4.6. Percentage of total U.S. input use by crop, in the Base

Crops	Acres	Nitrogen	Pesticide expenditures
	Input Type[a]		
Corn and Sorghum Grain	19	45	34
Barley, Oats, and Wheat	20	19	6
Cotton	2	4	15
Soybeans	25	2	36
Legume and Nonlegume Hay	8	6	1
Corn and Sorghum Silage	9	13	7

[a]Percentages do not sum to 100 because the table does not include all crops in the model.

Changes in Regional Crop Production Patterns

The changes in cropping practices required by the Soil Conservation Alternative as compared to the Base Alternative alter regional crop production patterns (Table 4.7). Small grain production increases substantially in the Corn Belt and the Delta States regions, offsetting the declining production of the row crops of corn, sorghum, and soybeans. This substitution of crops in the Corn Belt and the Delta States regions is needed because of erosion problems caused by rowcropping in these regions. The net effect of the soil erosion control policy in the southeastern United States is a shift away from rowcropping. Compared to the crop production patterns under the Base Alternative, corn production is increased (by means of practices that control erosion); however, its greater acreage is more than offset by a reduction in the acreages of soybeans and cotton. Because of the small erosion hazard in the Northern Plains, the production of corn, sorghum, and soybeans shifts to this region under the Soil Conservation Alternative. For the same reason, cotton production shifts from the Appalachian and Southeastern regions to the Western region.

Table 4.7. Changes in regional production by crop in each reporting region between the Soil Conservation Alternative and the Base (1,000 units)[a]

Region	Corn and Sorghum Grain	Barley, Oats, and Wheat	Oilmeal	Cotton	Legume and Nonlegume Hay	Corn and Sorghum Silage
	---------(bushels)--------		---(cwt.)---	--(bales)--	---------(tons)---------	
New England	-52,701	-23,388	5,813	0	845	-2,253
Appalacian	76,591	-32,503	12,250	-194	7,266	-5,738
Southeast	105,136	-6,390	-88,298	-1,734	2,654	2,115
Lake States	-12,972	25,064	20,669	0	-2,586	-972
Corn Belt	-428,347	601,824	-158,406	0	20,535	-53,786
Delta States	-87,921	226,614	-54,386	429	-1,019	-9,360
Northern Plains	229,327	97,434	141,455	0	2,908	-9,716
Southern Plains	-14,014	24,416	36,177	294	9,171	-66,274
Mountain	39,947	-13,038	17	0	2,536	3,007
Southwest	10,447	-5,771	154	50	603	-550
Northwest	-69,938	62,644	7	0	-257	-268
Pacific	4,135	-25,594	3,463	1,155	-458	-1,787
United States	-198,309	899,655	-81,082	0	41,482	-145,582

[a]Positive (negative) values indicate an increase (decrease) in production of the crop in the Soil Conservation Alternative compared to the Base.

Shifts in regional crop production and the changes in the soil management practices result in substantial decreases in rates of soil erosion (Table 4.8). The Southeastern and the Corn Belt regions, with their high rates of erosion, greater rainfall, and more sloping land, are affected most.

Table 4.8. Comparison of average rates of soil erosion by reporting region in the Base and the Soil Conservation Alternatives

Regions	Annual Soil Loss: Base	Annual Soil Loss: Soil Conservation Alternative
	(tons/acre/year)	
New England	5.9	2.0
Appalachian	5.5	2.2
Southeast	11.0	2.5
Lake States	3.3	2.2
Corn Belt	8.8	2.7
Delta States	5.2	3.1
Northern Plains	1.1	1.3
Southern Plains	3.7	2.1
Mountain	4.6	2.1
Southwest	.9	1.0
Northwest	1.9	1.5
Pacific	.8	.7
United States	5.0	2.1

Beef cattle replace hogs to an extent in the Corn Belt because of the substitution of hay for corn production. Most of the displaced hog production shifts to the Northern Plains because of the region's increased feed grain production in the Soil Conservation Alternative. The size of the beef cattle industry declines in the Northern Plains as production shifts elsewhere. Both beef cattle and hog production increase in the Appalachian region.

Commodity Prices and Regional Incomes

The shifting of crop production to less productive lands and the use of costly practices to control soil erosion raises the supply prices (shadow prices) for the commodities in the Soil Conservation Alternative (Table 4.9) from their level in the Base Alternative. Soybeans are especially affected by the soil conservation restraint. Their supply price is 43 percent greater under the Soil Conservation Alternative than under the Base Alternative. The supply price of wheat, however, declines by 3 percent as production shifts to the more productive lands of the midwestern United States. These higher supply prices for the feedstuffs, in turn, cause higher prices for the livestock products.

The higher commodity prices, in combination with the interregional shifts of crop and livestock production, result in a moderate increase in the total value of agricultural output in the Corn Belt and Lake States regions and a substantial increase in the Appalachian and Northern Plains regions (Table 4.10).

Table 4.9. Percentage change in the supply prices for the commodities in the Soil Conservation Alternative compared to the Base

Commodities	Percentage Change	Commodities	Percentage Change
Crops			
Corn	7	Soybeans	43
Sorghum	1	Legume hay	5
Barley	3	Nonlegume hay	3
Oats	12	Silage	3
Wheat	-3	Cotton	9
Livestock products			
Beef	5		
Milk	2		
Pork	7		

Table 4.10. Percentage changes in the regional value of agricultural production in the Soil Conservation Alternative relative to the Base

Regions	Percentage Change	Regions	Percentage Change
New England	-4	Northern Plains	30
Appalachian	28	Southern Plains	4
Southeast	-3	Mountain	19
Lakes States	7	Southwest	2
Corn Belt	8	Northwest	7
Delta States	-1	Pacific	3

Land Values

The results from the analysis imply differential capital gains and losses for landowners. As a national average, the imputed return to land subject to excessive erosion declines because of both the additional expense of controlling soil erosion and shifts to less intensive cropping systems (Table 4.11). The return to land most susceptible to erosion (Group 4) declines by 3 percent. The return to land least susceptible to erosion (Group 1) increases by 32 percent in the Soil Conservation Alternative compared to that in the Base Alternative. These changes in land returns result in a significant redistribution of regional asset values. Higher returns to land occur in the Appalachian, Lake States, and Delta States regions. Regions that have reductions in land returns as a result of the soil conservation policy are the New England, Southeast, Southern Plains, and Northwest regions.

Table 4.11. Percentage change in the regional imputed returns to land in the Soil Conservation Alternative relative to the Base

Region	Percentage Change	Region	Percentage Change
New England	-5	Northern Plains	25
Appalachian	109	Southern Plains	-6
Southeast	-43	Mountain	24
Lakes States	71	Southwest	17
Corn Belt	35	Northwest	-10
Delta States	57	Pacific	27

ANALYSIS OF THE NITROGEN RESTRICTION ALTERNATIVE

The Nitrogen Restriction Alternative restrains nitrogen use to a maximum of 50 pounds per acre. Hence, yields decline for crops using more than 50 pounds per acre in the Base Alternative. These yield changes are the only difference between the two alternatives. The yield changes are specified using the Spillman production functions estimated for each crop by producing area (Meister and Nicol, 1975).

The result from the Nitrogen Restriction Alternative is that the U.S. agricultural sector can meet the projected domestic and foreign demands when fertilizer rates are restricted. Comparison of the Base Alternative with the Nitrogen Restriction Alternative indicates the extent to which the lower yields force a substitution of land and other resources for nitrogen and alter regional production patterns. These changes and the implications for regional incomes and land values are briefly discussed in the following sections.

Resource Use

When the use of nitrogen is limited in the Nitrogen Restriction Alternative, other inputs must be substituted so that agriculture can continue to meet the given food and fiber demands (Table 4.12). For example, the restriction on nitrogen application rates lowers average cotton yields by 25 percent. Because of the reduced cotton yields, agriculture is forced to use more land (28 percent) and pesticides (35 percent) to maintain production levels.

Table 4.12. Percentage change of output and input by commodity group in the Nitrogen Restriction Alternative compared to the Base

	Change from the Base[a]			
Commodity groups	Production	Acres	Nitrogen	Pesticide expenditures
	(percent)			
Corn and Sorghum Grain	-4	14	-47	7
Barley, Oats, and Wheat	8	12	-3	26
Cotton	0	28	-56	35
Soybeans	-1	-1	54	-1
Legume and Nonlegume Hay	6	16	-44	22
Corn and Sorghum Silage	-6	2	-23	-2

[a]A positive (negative) value indicates a percentage increase (decrease) in the Nitrogen Alternative compared to the Base. A breakdown of resource use in the Base is provided in Table 4.6.

The differential changes in crop yields resulting from the nitrogen restriction alter the relative profitability of producing various crops. The result is a different mix of crops in the Nitrogen Restriction Alternative than in the Base Alternative. The production of corn and sorghum declines while the output of small grains and hay increases. Because of differences in input requirements for these crops, total resource use in agriculture also changes. For the agricultural sector as a whole, limits on the use of nitrogen force 25 million additional acres into crop production and raise total pesticide expenditures by 8 percent. In general, the 25 million acres are the result of small increases in each of the reporting regions, except for the Northern Plains region. This region increases the use of its available cropland by 23 percent.

In addition to requiring more land and other inputs to maintain agricultural output, the lower yields alter regional production patterns. Corn production decreases in the Corn Belt and the Appalachian regions while small grain, hay, and silage production increases. In response to the changed crop mix, beef cattle are substituted partially for hogs in the Corn Belt and Appalachian regions. Some of this hog production shifts to the Lake States and Northern Plains regions.

Soil Erosion

Soil erosion increases when agriculture is forced to bring additional land into production as a substitute for nitrogen. Nationally, total soil erosion is 14 percent higher in the Nitrogen Restriction Alternative than in the Base Alternative. Erosion increases as more land is brought into production and the use of intensive rowcropping systems on land already in production increases.

Regional Incomes and Land Values

The total value of agricultural production changes substantially in three of the reporting regions when the Nitrogen Restriction Alternative is compared to the Base Alternative (Table 4.13). The total value of production in the Lake States increases due to expanded swine production. Increases in the production of hogs, corn, and soybeans account for higher total value of output in the Northern Plains. A drop in the total value of production in the southwestern United States occurs because of a decline in feed production, and hence cattle feeding, under the Nitrogen Restriction Alternative. The total value of production in the Lake States increases due to expanded swine production. Increases in the production of hogs, corn, and soybeans account for higher total value of output in the Northern Plains. A drop in the total value of production in the southwestern United States occurs because of a decline in feed production, and hence cattle feeding, under Nitrogen Restriction Alternative.

Table 4.13. Percentage changes in the regional value of agricultural production in the Nitrogen Restriction Alternative relative to the Base

Region	Percentage Change	Region	Percentage Change
New England	-6	Northern Plains	19
Appalachian	8	Southern Plains	3
Southeast	7	Mountain	4
Lake States	10	Southwest	-32
Corn Belt	-1	Northwest	8
Delta States	3	Pacific	7

The Nitrogen Restriction Alternative also implies capital gains for some landowners and losses for others. As per acre production declines and crop production patterns are altered, the returns imputed to land change (Table 4.14). For example, the Southwestern and the Pacific regions have lower land values. Because these regions are intensive users of nitrogen for irrigated crops, the nitrogen restriction heavily impacts both net income and resource values.

Table 4.14. Percentage changes in the regional returns to land in the Nitrogen Restriction Alternative relative to the Base

Region	Percentage Change	Region	Percentage Change
New England	46	Northern Plains	9
Appalachian	18	Southern Plains	8
Southeast	28	Mountain	5
Lake States	5	Southwest	-18
Corn Belt	5	Northwest	8
Delta States	4	Pacific	-15

ANALYSIS OF THE INSECTICIDE RESTRICTION ALTERNATIVE

The potential impact of a ban on the use of Chlordane and Heptachlor is analyzed in this section. The Base Alternative allows corn producers in the Midwest to use these insecticides to protect their crop from soil insects. The Insecticide Restriction Alternative is formulated by selecting substitutes for these two insecticides and then making appropriate adjustments for costs and yields in the model's cropping activities. The substitutes selected are Thimet, Mocap, Dasanit, and Furadan. (The cost and yield adjustments were made with the help of Dr. H. J. Stockdale and Dr. G. R. DeWitt of the Iowa State University Department of Entomology.) These materials are more expensive than Chlordane and Heptachlor and equally effective except for two insect problems: the first-year insect complex of wireworms and grubs in corn following a grass crop, and cutworm damage to corn grown in lowland areas.

To reflect the relative ineffectiveness of the substitutes with the first-year insect complex, it is assumed that 20 percent of first-year corn suffers a 5 percent yield reduction. The net result is a 1 percent yield reduction on an average acre of first-year corn following a grass crop.

The ineffectiveness of the substitutes against cutworms also requires changes in yields and costs. It is assumed that 15 percent of the lowlands in land quality group 2 will be attacked by cutworms and that 25 percent of these infested acres will have to be replanted. For the other 75 percent of the infested acres it is assumed that three-quarters will receive additional insecticide applications as a rescue treatment but still suffer a 15 percent yield reduction, and the other one-quarter will be untreated and suffer a 25 percent reduction.

The results indicate few major changes in total national resource use in agriculture or in the supply prices of commodities, including corn, under the Insecticide Restriction Alternative. In the Corn Belt region, corn production declines about 10 percent, while the acreage of soybeans and small grains increases by 3 and 4 percent, respectively. However, these national adjustments do not account for the losses to be incurred by some corn producers. On the average, the crop losses are small, but because insect damage may range from zero to a total loss, the incomes of some farmers may be significantly reduced by a ban on Heptachlor and Chlordane.

ANALYSIS OF THE FEEDLOT RUNOFF CONTROL ALTERNATIVE

Requiring feedlot operators to control the runoff from their feedlots to prevent the pollution of nearby waterways raises the cost of livestock production. The Feedlot Runoff Control Alternative is analyzed to determine if comparative production advantages are altered enough by this regulation to cause regional shifts in livestock production.

To prevent livestock wastes from being washed into nearby waterways requires the construction of runoff control facilities for the feedlot. Control facilities as specified in this analysis include a diversion dam to route clean water from surrounding areas away from the feedlot, a sediment basin to separate the solids from the runoff for later disposal on the land, and a lagoon to impound the feedlot runoff for disposal by evaporation or irrigation.

Budgets have been developed showing the added expense to livestock producers who will be required to construct these facilities (Buxton and Ziegler, 1975; Johnson et al., 1975; Van Arsdall et al., 1974). Adjusting these budgets to reflect regional differences in climate, size of livestock enterprises, and the proportion of livestock in feedlots whose runoff may enter a waterway gives the annual cost of production increases for the livestock activities in the model. The annual cost of runoff control by type of livestock is calculated as:

$$C_{ik} = \sum_{j} RC_{jik} PM_{jik} P_{jik} \tag{4.1}$$

j = 1, ..., n for livestock enterprise sizes;
i = 1, ..., 28 for the market regions;
k = 1, 2, 3 for livestock types–beef feeding, dairy, and hogs.

where C_{ik} = the annual cost of runoff control for a representative animal of type k in market region i; RC_{jik} = the annual cost for runoff control for a representative animal of type k in lot size j in market region i; PM_{jik} = the proportion of total animals of type k in market region i marketed from lot size j; P_{jik} = the estimated proportion of feedlots of size j in market region i with animals of type k whose runoff may enter a waterway. These values are added to the nonfeed costs in the model to create the Feedlot Runoff Control Alternative.

When these costs are included in the model for the Feedlot Runoff Alternative, there is a slight shift of beef cattle from the Lake States to the Corn Belt and from the Northern Plains to the Southern Plains. There also is a small shift of hog production from the Corn Belt to the Northern Plains.

Comparison of the results from the Feedlot Runoff Control Alternative with those of the Base Alternative indicates few important changes in total resource use in agriculture or in the supply prices of commodities, including beef and pork. The small increase in the shadow price of livestock products does not mean that all livestock producers would be unaffected. Because of the expense for runoff control facilities, farmers will be earning a lower rate of return than expected on their investments in feedlot facilities. Small operators would be most affected because the cost of runoff control facilities increases sharply with decreasing lot size.

ANALYSIS OF THE EXPORT POTENTIAL ALTERNATIVES

This section provides an analysis of the impact of higher production costs caused by environmental policies on the potential export capacity of U.S. agriculture. Higher production costs decrease the export capacity of agriculture for a given export price because marginal land, formerly profitable to crop, will be taken out of production.

The first of the two high-export alternatives, the High Export Alternative, allows production to expand until the aggregate supply price for corn, wheat, oilmeals, and sorghum is equal to an export price allowing almost all available cropland to be brought into production. (The four commodities are exported in fixed proportions to prevent specialization.) This expansion occurs without any controls on the environmental consequences of the increased production. The second export alternative, the Restricted Export Alternative, has the same export price as the High Export Alternative but incorporates the restrictions associated with all four of the environmental policies analyzed earlier. Comparison of the results from these two export alternatives provides a measure of the extent to which the higher production costs resulting from the imposition of the environmental policies lowers the potential export capacity of agriculture.

High Export Alternative

The High Export Alternative uses 67 million more acres than does the Base Alternative. In addition, the expansion of exports shown in Table 4.15 requires more fertilizer and pesticides. The High Export Alternative uses 29 percent more nitrogen and increases pesticide expenditures by 50 percent. Most of the nitrogen increase is due to the high requirements of corn and sorghum. The largest

Table 4.15. Comparison of export levels for the Base, High Export, and Restricted Export Alternatives

Commodity	Alternative		
	Base	High Export	Restricted Export
	(thousand tons)		
Corn	27,692	60,844	37,005
Wheat	23,220	37,764	27,306
Oilmeal	22,562	52,406	30,946
Sorghum	4,480	9,255	5,821

proportion of the increase in pesticide expenditures is for corn, sorghum, and soybeans.

Regional crop production patterns are stable except for a relatively large increase of corn and sorghum in the Northern Plains and an increase in the concentration of cotton production in the Delta States region. There are some regional shifts in livestock production compared with the Base Alternative. The increase in corn and sorghum production in the Northern Plains favors swine production, as hogs shift from the Corn Belt to the Northern Plains. Beef feeding partially replaces hogs in the Corn Belt region. Beef cattle displaced by hogs in the Northern Plains shift to the Southern Plains.

The soil management practices change in the High Export Alternative relative to the Base Alternative. Continuous rowcropping increases as the production of corn, sorghum, and soybeans for export expands. The number of acres protected by strip cropping and terracing is higher because of the increased rowcropping of land especially susceptible to soil erosion. Because of the large increase in cultivated acres not adequately protected by soil conservation practices, total soil erosion increases by 21 percent in the High Export Alternative.

Restricted Export Alternative

The reduced export capacity of the Restricted Export Alternative relative to the High Export Alternative is due partly to reduced land utilization, since cropland having severe soil erosion problems is not cropped. The extent to which cropland is idled in the Restricted Export Alternative is further increased by the nitrogen restriction. The nitrogen restriction reduces crop yields to such an extent that many acres of marginal land cannot produce enough to cover the cost of the required soil conservation practices. In addition to idling land, the environmental policies alter regional cropping patterns. There is a considerable shift of corn, sorghum, and soybean production from the Corn Belt to the Northern Plains, where fewer erosion problems exist. These crops replace the small grains produced in the Northern Plains. Some small grain production shifts to the Corn Belt as less intensive rowcropping rotations are used to reduce erosion losses.

Some hog production shifts from the Corn Belt to the Northern Plains and the Lake States consistent with the regional changes in corn production. Both the Corn Belt and the Delta States regions feed fewer cattle because of the erosion hazard of growing the corn grain and corn silage to feed them. These displaced feeders are dispersed across the United States, with the largest number going to the Pacific region.

Soil erosion declines by 49 percent from the level in the Base Alternative, even though 55 million additional acres are cropped. This significant decline in soil erosion occurs because of expanded use of rotations with hay and small grains, and the increased use of strip cropping, terracing, and reduced tillage to protect the soil.

High Export Alternative Compared to the Restricted Export Alternative

Imposing the environmental restraints on agriculture in the Restricted Export Alternative reduces the dollar value of exports by 40 percent compared to the High Export Alternative. These environmental restraints make crop production unprofitable on 12.6 million acres of available cropland and cause a 25 percent decline in land returns.

GENERAL IMPACT

All facets of an environmental improvement program stand to have differing effects on U.S. agriculture. The nation has sufficient producing capacity so that modest restraints on supply exercised through an environmental policy can allow domestic and export demands to be met without large increases in commodity supply prices. Since the United States is a large exporter of grains and cotton, supplies for domestic consumption are not endangered by any of the alternatives evaluated. Under each environmental alternative examined, some regions gain while others sacrifice in income and asset values. Regions that specialize in row crops, such as corn, cotton, and soybeans, and that have sloping land and heavy rainfall would sacrifice under restraints on soil erosion and fertilizer use.

REFERENCES

Buxton, B. M., and S. J. Ziegler. 1974. Economic Impact of Controlling Surface Water Runoff from U.S. Dairy Farms. Agricultural Economic Report 260. Economic Research Service, USDA, Washington, D.C.

Conservation Needs Inventory Committee. 1971. National Inventory of Soil and Water Conservation Needs 1967. USDA Statistical Bulletin 461. Washington, D.C.

Johnson, J. B., G. A. Davis, J. R. Martin, and C. K. Gee. 1975. Economic Impact of Controlling Surface Water Pollution from Fed Beef Operations. Agricultural Economic Report 292. Economic Research Service, USDA, Washington, D.C.

Meister, A. D., and K. J. Nicol. 1975. A Documentation of the National Water Assessment Model of Regional Agricultural Projections, Land and Water Use and Environmental Interaction. Center for Agricultural and Rural Development, Iowa State University, Ames.

Van Arsdall, R. N., R. B. Smith, and T. A. Stucker. 1974. Economic Impact of Controlling Surface Water Pollution from U.S. Hog Production. Agricultural Economic Report 263. Economic Research Service, USDA, Washington, D.C.

Vocke, G. F., E. O. Heady, W. G. Boggess, and H. J. Stockdale. 1977. Economic and Environmental Impacts on U.S. Agriculture from Insecticide, Fertilizer, Soil Loss, and Animal Waste Regulatory Policies. CARD Report 73. Center for Agricultural and Rural Development, Iowa State University, Ames.

Wischmeier, W. H., and D. D. Smith. 1965. Predicting Rainfall-Erosion Losses from Cropland East of the Rocky Mountains. USDA Agriculture Handbook 282. Washington, D.C.

CHAPTER 5. A STUDY OF THE TRADEOFFS BETWEEN SOIL EROSION CONTROL AND THE COST OF PRODUCING THE NATION'S AGRICULTURAL OUTPUT

by Orhan Saygideger and Earl O. Heady

THE MAJOR OBJECTIVE of this study is to estimate the tradeoff rate between various levels of soil erosion control in the U.S. agricultural sector and the cost of producing the nation's agricultural output. The analysis is made by means of an interregional competition linear programming model of U.S. agriculture incorporating a multigoal objective function. This multigoal objective function allows explicit consideration of the tradeoffs between the goals of soil conservation and the cost of producing the nation's food supply.

The model employed in the analysis is the same as in Chapter 4, except for the objective function. The remainder of this chapter presents a very brief description of the model (the interested reader may refer to Chapters 3 and 4 for details) and the results of the analysis.

FORMULATION OF THE STUDY USING THE PRIOR WEIGHTING TECHNIQUE

For this study the goals of minimizing production and transportation costs and improving soil conservation are combined into a single objective function with the assignment of explicit weights to each goal. The *a priori* specification of weights indicating the relative importance of each goal yields a composite linear objective function. The tradeoff between the goals can be generated by repeatedly optimizing this function with the model as the weights are parametrically varied.

Mathematical Description of the Multigoal Problem

The general mathematical description of the multigoal problem with p goals can be specified as follows:

$$\text{Min } F = Cx \tag{5.1}$$

$$\text{Subject to } Ax \leq b \qquad x \geq 0 \tag{5.2}$$

where F is a p × 1 vector; C is an f × n matrix; and x is an n × 1 vector. A is an m × n matrix, and b is an m × 1 vector. Two goals are evaluated in this study and equation (5.1) thus is:

$$\text{Min } F = (F_1(x), F_2(x))^T = Cx \tag{5.3}$$

where C is now a 2 × n matrix. The generation of a set of solutions for the tradeoff analysis can be obtained by transforming the vector-valued objective function in equation (5.3) into a scalar-valued function in the following manner:

$$\text{Min } \sum_{i=1}^{2} w_i f_i(x) \tag{5.4}$$

where the w_i's are the relative weights assigned to each objective (all $w_i > 0$). Systematically varying the w_i's in equation (5.4) will allow a series of solutions of the programming model, which can yield a tradeoff curve between the goals.

Specification of the Model

The model is defined with sets of regions consistent with land and water resources, crop and livestock production possibilities, and interregional interaction of U.S. agriculture expected in the year 1985 (Saygideger et al., 1977). In the specification of the model, the land resources of U.S. agriculture are defined in 5 land quality groups for each of 105 producing areas representing homogeneous production conditions as explained in Chapters 3 and 4. Contiguous producing areas are explained in Chapters 3 and 4. Contiguous producing areas are aggregated to form the 28 market regions in Figure 3.2. Crop production activities, including tons of soil eroded per acre per year, are defined by land quality group in each producing area. Endogenous livestock production activities for swine, dairy, beef feeding, and beef cows are defined in each market region. All other livestock, including poultry, are exogenous to the model. All production activities are based on technologies projected to be available in 1985. Production costs include market rates of return to all resources used in agriculture except land and water. Land and water returns are determined endogenously in the model. Fixed commodity requirements are defined by market region based on projections of population, per capita incomes, and exports for 1985. The model also incorporates a commodity transportation subsector among consuming regions to allow interregional shipment of commodities.

The choice of weights in equation (5.4) indicates the relative importance of the goals. Thus the technique can be used for objective functions defined in different units, such as production costs in dollars and pollution measured in eroded soil. As Candler (1973) indicates, one of the goals can be given a weight of 1.0. Then the other goal weight has significance relative to this "numerary" goal. This study defines the cost goal as the numerary goal. Hence the weights assigned the soil-loss goal can be interpreted as the pollution cost of a ton of eroded soil to society. By systematically varying the weights assigned the soil-loss goal, the study traces the tradeoffs between the two goals.

The schematic in Figure 5.1 illustrates the framework for a hypothetical multigoal model with two producing areas aggregated to form one market region. The schematic also shows how the objective function interacts with the rest of the model via the goal-accounting restraints.

	Relationship	Crop Production in PR 1	Crop Production in PR 2	Livestock Production in MR 1	Cost Goal	Soil-Loss Goal
Objective Function					w_1^a	w_2
Cost Accounting Restraint	0 =	VC^b	VC	VC	-1	
Soil Loss Accounting Restraint	0 =	SL^c	SL			-1
Resource Restraints in PA 1	$\geq$	B_{11}^d				
Resource Restraints in PA 2	$\geq$		B_{21}			
Commodity Demands in MR 1	$\leq$	C_{11}^d	C_{21}			

[a] w - weights assigned to the goal activities.
[b] VC - variable costs of production.
[c] SL - soil loss in tons/acre.
[d] B_{ij}, C_{ij} - interaction coefficients for PA_i in MR_j.

Figure 5.1. Illustrative framework for a hypothetical multigoal model with two producing areas (PA) aggregated to form one market region (MR).

RESULTS

To generate the tradeoff curve in Figure 5.2, six solutions for the model are obtained, each with different relative weights on the two goals incorporated in the model. Solution 1 has a weight of $1.00 for the cost-of-production goal and zero for the soil-loss goal. In Solutions 2, 3, 4, and 5 the weight on the cost-of-production goal is kept at $1.00, but the weights on the soil-loss goal are $2.50, $5.00, $10.00, and $20.00, respectively. As the weighting of the soil-loss goal increases, the penalty on soil erosion increases. For Solution 6, the cost-of-production goal has a zero weight while the soil-loss goal has a weight of $20.00. Because each solution is an efficient point between the two goals, these points can be plotted and used to draw the tradeoff curve between the goals. The shape of this tradeoff curve, as indicated in Figure 5.2, implies that society may need to make a sizable sacrifice in one goal in order to minimize the other goal taken alone. If society is interested only in minimizing production costs in U.S. agriculture (Point 1 in Figure 5.2), the result is high rates of soil erosion from U.S. cropland. Conversely, if a high level of soil conservation alone is desired (Point 6 in Figure 5.2), then minimizing only the soil-loss goal results in substantial increases in the cost of production. The intermediate solutions indicate a "corner" for the tradeoff curve between the goals.

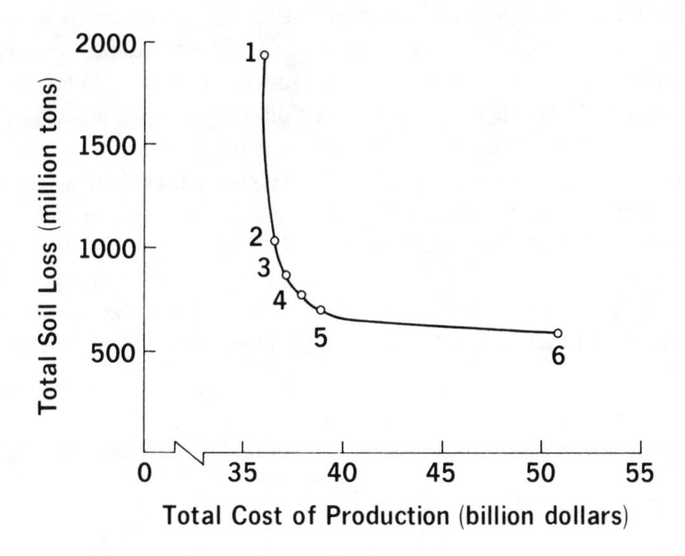

Figure 5.2. Trade-off frontier for the two goals of production costs and soil-loss control, U.S. totals.

National and Regional Changes in Production Patterns

For each set of relative values attached to the production costs and soil-loss goals, the model selects regional patterns of production consistent with each region's comparative advantage in producing crop and livestock products. When a zero value is placed on the soil-loss goal, there is no penalty for erosion. Crops are grown in regions that have a comparative advantage for them in terms of yields, production costs, location, and transportation costs. When positive values are assigned to the soil-loss goal, the model selects the regional levels and combinations of crop and livestock production that have comparative advantage in meeting both goals.

The total acres cropped varies less than 2 percent between alternative solutions as the level of soil conservation increases due to heavier penalties on soil erosion losses. The cropping of total available farmland in the United States declines only slightly in Solution 2 compared to Solution 1, then rises in Solutions 3, 4, 5. (See Table 5.1.) In Solution 2, the assigned cost of $2.50 per ton of soil eroded results in an optimal organization for U.S. agriculture that includes a greater use of reduced tillage and terracing relative to Solution 1 (Figures 5.3 and 5.4). Average crop yields increase because the greater use of these management practices more than offsets any crop yield declines caused by regional shifts of production to areas of lesser productivity. With the higher soil-loss penalties in Solutions 3, 4, and 5, however, the interregional adjustments of production to areas of lesser productivity outweigh any gains resulting from improved management practices and average yields decline.

Regional utilization of available cropland does vary significantly as the penalty for soil erosion losses increases. In particular, the South Atlantic region shows a significant decline in the percentage of cropland in cultivation. The Great Plains region shows a significant increase. These regional results are associated with the relative erosion hazards in these regions. When a relatively high penalty is placed on eroded soil, those regions with cropland more susceptible to erosion than other regions are placed at a comparative disadvantage, and the model shifts crop production to regions with fewer erosion hazards. The cropping pattern of the land that is cultivated changes in response to placing a penalty on soil erosion losses. At the national level, total row-crop acreage declines while the acreages of small grains and hays increase when Solution 1 is compared to the other solutions

Table 5.1. Percentage of available cropland utilized by region

Regions	Solutions 1	2	3	4	5
United States	94	93	94	94	95
North Atlantic	98	94	93	95	95
South Atlantic	96	94	94	89	89
North Central	96	96	97	97	98
South Central	95	95	92	93	94
Great Plains	90	88	93	93	94
Northwest	91	86	86	87	96
Southwest	92	87	88	95	96

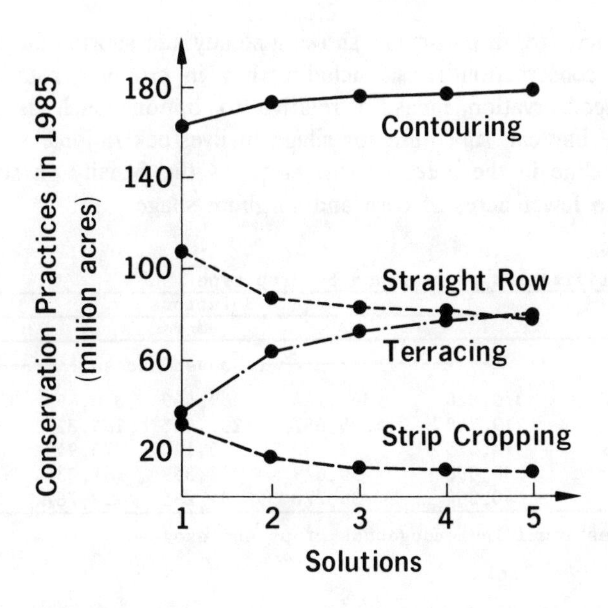

Figure 5.3. Changes in acres under conservation practices with alternative solutions.

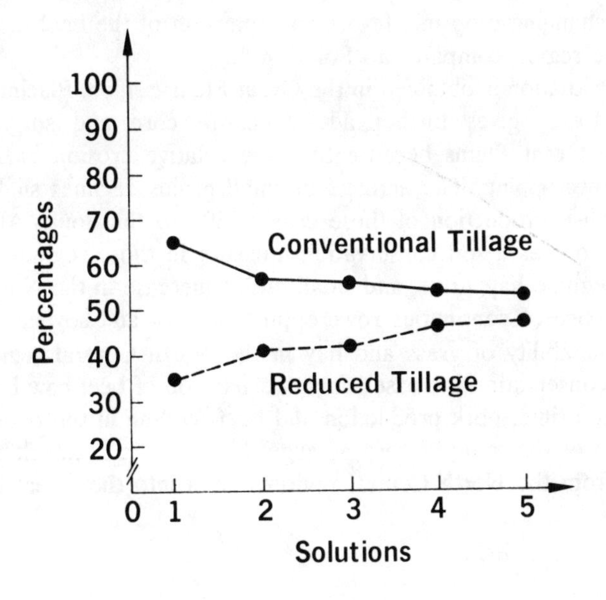

Figure 5.4. Changes in proportions of two tillage practices utilized under alternative solutions in 1985.

(Table 5.2). Hay acreage, in particular, shows a steady and substantial increase as the level of soil conservation rises. Including hay in cropping rotations is an economical soil conservation measure relative to options such as additional terracing, and the hay can substitute for silage in livestock rations. A significant portion of the decline in the acres of row crops as the penalty on soil erosion increases is due to fewer acres of corn and sorghum silage.

Table 5.2. Utilization of cropland by crop type

Crop Type	Solutions 1	2	3	4	5
	(thousand acres)				
Total cropland	370,826	366,144	369,469	370,468	373,974
Row crops	219,749	205,657	201,685	199,823	202,311
Small grains	72,675	73,530	75,140	75,944	75,333
Hays	38,098	50,679	58,359	61,934	65,070
Others[a]	40,304	36,278	34,285	32,767	31,290

[a]Includes remaining endogenous crops and exogenous crops.

Assigning a penalty per ton of soil eroded significantly alters regional comparative advantages for crop production. The high erosion hazard associated with rowcropping in the South Atlantic region results in a substantial shift of soybean and cotton production away from this region. Legume hay, grass, and small grains substitute for these crops because of the protection they provide for the topsoil. This changing crop mix favors an expansion of the beef cow herds in the South Atlantic region compared to Solution 1.

The opposite situation is obtained in the Great Plains region. Placing a penalty on soil erosion losses gives further advantages to corn and sorghum grain production in the Great Plains because they are relative erosion hazards when associated with rowcropping. The acreage of small grains declines slightly in the Great Plains because production of these crops shifts to the South Atlantic and North Central regions as a soil conservation measure in those regions.

Acreages of legume hay, grass, and small grains increase in the North Central region as the practice of continuous rowcropping of corn and soybeans declines. The increasing availability of grass and hay in the North Central region, as the emphasis on soil conservation increases, favors expansion of beef cow herds in the region. At the same time, pork production and beef feeding in the region decline somewhat because of the reduced corn acreage. Most of the livestock production that shifts away from the North Central region moves into the Great Plains and Southwest regions.

Changes in Farming Practices and Soil Losses

Imposing a penalty on soil erosion losses will force agriculture to make major adjustments in farming methods and cropping patterns and can significantly improve soil conservation. Reduced-tillage practices are substituted for conventional-tillage practices to increase the quantity of plant residues on the soil surface. Contour farming is substituted for straight-row farming on land with a relatively small erosion hazard, while terracing is used on those fields subject to more severe erosion problems. When minimizing only production and transportation costs, 33 percent of the nation's cropland is under straight-row farming. When a soil erosion penalty is added to production expenses, the percentage of cropland under straight-row farming declines to 23 percent of the total cropland in Solution 5. The acres of cropland protected by terracing increases from 11 percent of the total in Solution 1 to 23 percent in Solution 5 (Table 5.3).

Table 5.3. Average annual soil loss per acre on cultivated land

	Solutions				
Regions	1	2	3	4	5
	(tons per acre per year)				
North Atlantic	5.65	2.76	2.65	2.32	1.91
South Atlantic	12.58	6.61	5.62	4.05	3.31
North Central	4.80	3.07	2.79	2.67	2.39
South Central	4.77	2.76	1.64	1.59	1.51
Great Plains	4.68	1.76	1.29	1.27	1.29
Northwest	3.56	1.79	1.53	1.04	0.99
Southwest	1.29	0.96	0.85	0.81	0.74
United States	5.56	3.05	2.50	2.20	1.98

In general the changes in farming practices occurring between Solutions 1 and 2 are more significant than the changes from Solution 3 through Solution 5 because most practical adjustments occur between Solutions 1 and 2. This declining rate of change in farming practices as the penalty on soil erosion increases explains, in part, the corner on the tradeoff curve (Figure 5.2).

With no soil-loss penalty (Solution 1), average annual soil loss is 5.56 tons per acre (Table 5.4). (While this soil loss may seem relatively low, it should be remembered that it includes a large land acreage in which there is zero soil loss.)

Table 5.4. Levels of various production and environmental variables in each of the solutions[a]

Solution	Total Acres	Nitrogen Used	Pesticide Expenditures[b]	Reduced Tillage	Terracing	Average Annual Soil Loss
	(acres)	(tons)	(dollars)	(acres)	(acres)	(acre)
1	370,826	9,350	1,527,964	106,407	36,513	5.56
2	366,144	9,351	1,908,280	144,950	64,023	3.05
3	369,469	9,705	2,053,998	151,130	72,215	2.50
4	370,468	10,041	2,268,421	160,054	75,209	2.20
5	373,974	10,442	2,458,863	165,063	80,440	1.98

[a]In thousands of units.

[b]Pesticide expenditures are in 1972 dollars.

The average loss declines to 1.98 tons per acre as the penalty increases to $20 per ton. Nationally, total soil erosion drops 64 percent from Solution 1 to Solution 5. This decline in erosion is achieved partly by changing farming practices and partly by interregional adjustments in crop production patterns. The greatest reductions in soil erosion losses are obtained in the South Atlantic, Great Plains, and South Central regions. In these three regions, soil erosion declines by 74, 72, and 68 percent, respectively, when Solution 5 is compared to Solution 1.

Change in Use of Agricultural Inputs

The use of fertilizer and pesticides increases steadily as agriculture is reorganized in the model's solutions to provide more protection for the cropland (i.e., Solution 5 compared to Solution 1). Changing farm practices, such as the expanding pesticide requirements for crop production associated with the increased use of reduced tillage, significantly alters the use of inputs by U.S. agriculture (Table 5.3). The principal reason the use of fertilizer increases as the level of soil conservation improves is due to interregional adjustments in corn production. When agriculture is organized without consideration of the consequences of soil erosion, corn is grown on the most productive land in continuous row-crop rotations, especially in the North Central region. As the cost penalty assigned per ton of soil eroded rises, this pattern changes as hay, grass, and small grain crops are grown in rotation with the corn to control erosion. Thus, as corn production is forced onto less productive land within each region and to less productive regions, e.g., the Great Plains region, the amount of fertilizer and pesticides required, on the average, to grow a bushel of corn increases.

Supply Prices

Changes in farm practices such as the increased use of terracing, growing corn in rotation with grass and hay, shifting some of the corn acreage in the North Central region to the Great Plains, etc., cause only modest increases in the supply prices for crops up to Solution 3. However, between Solution 3 and Solution 5 supply prices increase by a large amount (Table 5.5). These large increases would raise food costs significantly for U.S. consumers and place U.S. agriculture at a disadvantage in world commodity markets.

Changes in Return to Land

Nationally, returns to land decrease only slightly in Solutions 2 and 3, compared to Solution 1, then increase sharply in succeeding solutions (Table 5.6). The decrease in net returns to land in Solution 2 can be attributed to increased production costs. Increased costs result as adjustments are made in cropping practices and as large interregional shifts in crop production occur between Solutions 1 and 2. Production costs also increase progressively between Solutions 2 and 5. However, supply prices of crops rise more rapidly than production costs.

Table 5.5. **Index of supply prices (shadow prices) for agricultural commodities (solution 1 = 100)**

Commodities	Solutions 1	2	3	4	5
	----------(percent change from solution)------				
Corn	100	104	115	144	198
Wheat	100	106	114	145	205
Soybeans	100	113	134	184	280
Hay	100	102	107	129	172
Cotton	100	92	104	115	136
Silage	100	102	109	132	185
Pork	100	105	113	133	174
Beef	100	101	107	123	155
Milk	100	102	106	116	137

Hence, in Solution 5 average returns to land increase from their level in Solution 1 by an average of about 150 percent for the whole United States. The increase in land returns exceeds 200 percent in the Great Plains and Southwest regions because these regions have large areas of level land not very susceptible to erosion. The increase in land returns is only 88 and 99 percent for the South Atlantic and South Central regions, respectively. The increase for these regions is low because of both more rainfall and large sloping areas of lands. These regions give up relatively more row crops and are forced to install relatively expensive conservation practices such as terraces as the value placed on soil losses increases.

Table 5.6. Index of return to land based on land shadow price (solution 1 = 100)

Regions	Solutions 1	2	3	4	5
	-------(percent change from solution)---------				
North Atlantic	100	97	104	148	240
South Atlantic	100	75	79	114	188
North Central	100	103	122	184	301
South Central	100	85	91	121	199
Great Plains	100	98	112	177	307
Northwest	100	107	114	182	297
Southwest	100	103	118	176	311
United States	100	96	109	162	268

POLICY IMPLICATIONS

The purpose of this study has been to provide information about the tradeoffs between the cost of producing and transporting agricultural products to current consumers and the value of preventing soil loss and maintaining a productive cropland base for future generations. The derivation of the tradeoff function between these two goals should provide policymakers with valuable information for decision making.

As presented in Figure 5.2, the points on this tradeoff curve show attainable combinations of total production costs and total soil erosion for U.S. agriculture. The determination of the optimal point on this tradeoff curve should depend on the preferences of decision makers representing society.

The shape of the tradeoff curve indicates that the costs of soil erosion abatement are not likely to vary proportionately to the amount of erosion abated. At high, unrestrained levels of soil loss, a given reduction in erosion can be obtained without substantial cost to society. When soil losses are at relatively low levels, however, further reductions are very expensive. In summary, the more soil loss is reduced on U.S. cropland, the more sharply costs will rise for further reductions.

Changes in farm practices required to abate soil erosion require new management skills and capital investments by farmers. In general, farms with lands susceptible to severe erosion and thus requiring costly conservation practices stand to be economically disadvantaged by soil conservation policies. Farmers with land not subject to severe erosion hazards can generate more income resulting in a higher capitalized value of land. A national program of erosion abatement also would redistribute income among regions. Regions of heavy rainfall and sloping lands are forced into less intensive agriculture and may have a sacrifice in farm income. Regions of moderate rainfall and level lands have the opportunity to be cropped more intensively, thus increasing the value of the land accordingly. These differential impacts should be recognized and considered in national policies directed at reducing soil erosion.

REFERENCES

Candler, W. 1973. Linear Programming in Capital Budgeting with Multiple Goals. In *Multiple Criteria Decision Making*, edited by L. J. Cohran and M. Zeleny. University of South Carolina Press, Columbia.

Saygideger, O. G., G. F. Vocke, and E. O. Heady. 1977. Analysis of Interaction between Soil Conservation and Agricultural Production in the United States Using a Multigoal Linear Programming Model. CARD Report 76. Center for Agricultural and Rural Development, Iowa State University, Ames.

SECTION III. EVALUATION FOR U.S. AGRICULTURE OF SEDIMENT-CONTROL PROGRAMS FOR THE NATION'S WATERWAYS

THE two interregional competition studies in this section address questions relating agricultural production practices directly to water quality in the nation's waterways. The main impact of agriculture on water quality is through sedimentation of the waterways. Sediment also transports some pesticides and phosphates into streams and lakes. The principal difference between these studies of water quality as affected by agriculture and the previous studies in this book is the addition of a sediment delivery subsector to the model. This subsector determines the sediment loads in the nation's waterways from the gross soil erosion in the crop sector of the model. This sediment delivery subsector allows for the investigation of the relationships between agricultural production practices on U.S. croplands and water quality (in terms of sediment loads).

The objectives of the first study are to formulate and quantify the sediment delivery system and then use the system to investigate the relationships and tradeoffs between river-basin sediment loads, agricultural production practices, and demand for food and fiber projected for the year 2000. The objective of the second study is to compare the consequences of controlling sediment delivered to the waterways by regulatory measures with control by a taxing scheme. Both studies utilize linear programming models based on the agricultural data base assembled for the National Water Assessment (NWA) study as described in Chapter 3.

The results in the first study of various alternatives for limiting sediment loads in streams from cropland indicate that agricultural adjustments in the Mississippi River drainage area offer the most potential for the control of soil erosion. The results from the second study imply there would be significant differences in the resource allocation obtained under the alternative sediment-control policies.

The taxing scheme to control sediment is more efficient and results in a less costly tradeoff between reductions in sediment loads in the nation's waterways and food cost increases.

CHAPTER 6. AN INTERREGIONAL MODEL FOR EVALUATING THE CONTROL OF SEDIMENT FROM AGRICULTURE

by James C. Wade and Earl O. Heady

AS a nonpoint source of pollution, sediment is not easily controlled. Land management systems must be imposed on large areas to reduce erosion and control sediment pollution.

Viewing sediment from eroded cropland and its associated chemical pollutants as a residual of cropland management, environmental quality improvement may be attained by changing the technologies applied to the land and the crops produced on it. These possible changes have many simultaneous interregional and national economic and physical implications. Changes in crop production in one region of the country to meet water quality demands require adjustments not only in that region but also in other regions. The study reported here has two major purposes. The first is to develop an interregional competition model of the U.S. agricultural sector that links agricultural practices on U.S. croplands to water quality through the delivery of eroded soil to the nation's waterways. The second purpose of the study is to apply this model to a set of hypothetical policy alternatives in the analysis of U.S. water quality problems related to sedimentation. The development of a sediment sector for the basic Center for Agricultural and Rural Development-National Water Assessment (CARD-NWA) model will be presented first, and then the results from the analyses of some policy alternatives will be discussed. For additional details about the CARD-NWA model, the reader is referred to Chapter 3 of this book. This study was carried out under a grant from the National Science Foundation's Research Applied to National Needs (RANN).

OVERVIEW OF THE MODEL

The agricultural sector model employed in this study is the CARD-NWA interregional competition model developed in Chapter 3. This model is expanded to include a sediment transport and delivery component that links gross soil erosion to sediment loads in the nation's waterways. Technology, population, export demands, and other relevant parameters are projected to the year 2000. The year 2000 provides sufficient time for regional agricultural adjustments to take place and thus allows for assessment of long-run effects of environmental policy. The agricultural sector of this model will be briefly discussed here, and then the development of the sediment sector will be presented in detail.

The linear programming model is subdivided into 105 agricultural producing areas which cover the entire United States. Production activities are defined in these producing areas to produce commodities for the 28 market regions in the model. The market regions are aggregations of contiguous producing areas. Final demands for commodities are established for each market region (including international exports), and transportation of most commodities occurs among the market regions. Thus, the economic sector of this spatial model uses resources, primarily land and water, to produce crop and livestock commodities and residual soil loss.

In this model, production of the major crops occurs in each of the 105 producing areas with the sediment direction flows shown in Figure 6.1. Each producing area has 9 land quality groups and is provided with linear programming activities that produce the endogenous crops in the model. The irrigated areas have 18 land groups. The endogenous crops are barley, corn grain, corn silage, cotton, legume hays, nonlegume hays, oats, sorghum grain, sorghum silage, soybeans, sugar beets, and wheat. Resources for the production of all other crops are preallocated in each producing area, i.e., these crops are exogenous to the

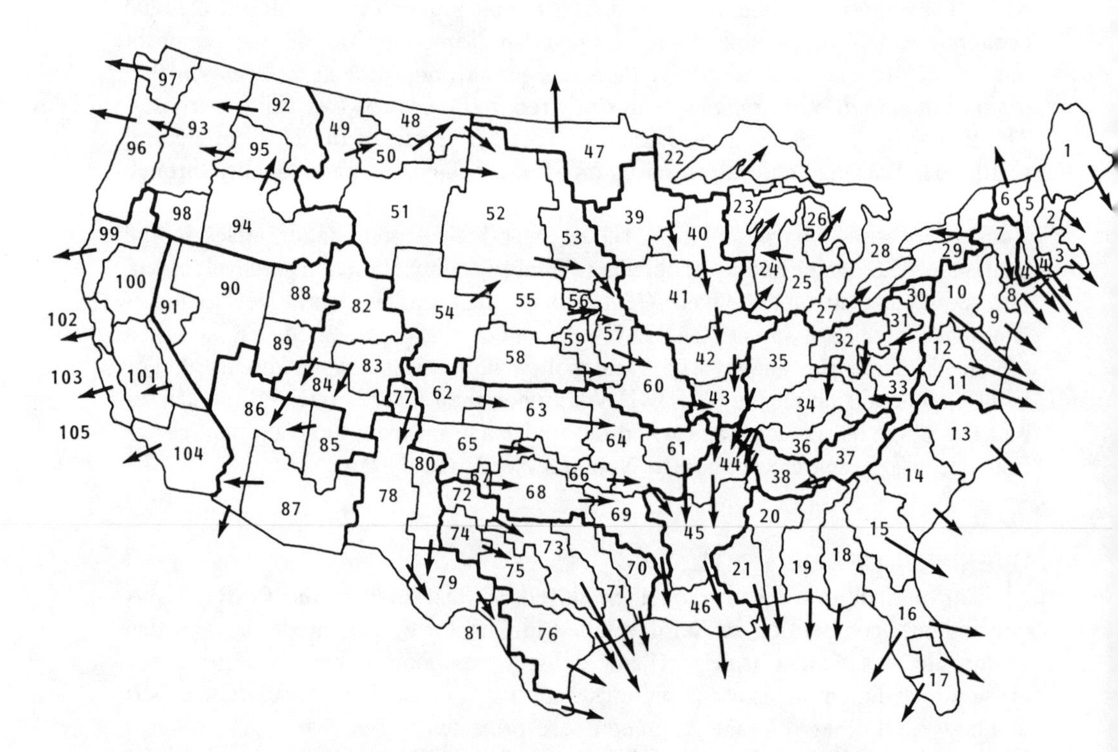

Figure 6.1. Producing areas (PA), river basins, and schematic river flows for the sediment transport submodel.

model. The crop commodities produced by these activities are shipped interregionally to meet the demands of each of 28 market regions shown in Figure 3.2. The total national cost of production and transportation of demanded commodities is minimized with resources receiving market rates of returns. Production technologies available on each soil group and in each producing area are the same as explained in Chapter 4. Each cropping activity produces a residual soil loss based on regional climatic and soil conditions. The soil loss for each activity is computed using the Universal Soil-Loss Equation (USLE) illustrated in equation (1.2).

Endogenous livestock production, defined at the market-region level, includes dairy, hogs, beef cows, and beef feeding, which provide five commodities (milk, pork, beef feeders, fed beef, and nonfed beef) for final demand. Various ration alternatives provided in the model are used in each market region to convert feed inputs into livestock outputs. Exogenous livestock production is predetermined for each market region from historical patterns and the feed requirements added to the respective market region commodity demands.

All technologies and feed conversion rates for crop and livestock production are projected to the year 2000. Costs are based on 1972 dollars.

Commodity demands for each market region are based on a projected national population of 264.3 million in the year 2000. Regional distribution of population and export levels are set as in the National Water Assessment model (Meister and Nicol, 1975; U.S. Water Resources Council, 1974). The endogenous commodities are shipped among market regions to meet consumer and export demands, except for cotton lint and beet sugar production, which are equated to national demands. Hay and silage are consumed only within the market region of production as interregional transportation of these commodities is excluded.

The basic model is summarized in matrix relationships (6.1)-(6.8):

$$\min Z = c'x \tag{6.1}$$

subject to

$$Ax \leq b \tag{6.2}$$

$$Bx \geq d \tag{6.3}$$

$$Sx - Dx_D \leq s^* \tag{6.4}$$

$$x_D - Tx_T 0 \tag{6.5}$$

$$x \geq 0 \tag{6.6}$$

$$0 \leq x_D \leq r_D \tag{6.7}$$

$$0 \leq X_T \leq r_T \tag{6.8}$$

where Z = the total production and transport cost of endogenous U.S. agricultural commodities; x = the agricultural commodity production and transport vector; c = the production and transport costs vector; A = the resource requirements matrix for the production and transport vector; b = the resource limits vector; d = final commodity demands vector; B = the commodity output matrix; S = the soil-loss or residual output matrix; x_D = the sediment delivery vector; D = the sediment transport ratio vector; r_D = the restraint vector on sediment delivery; r_T = the restraint vector on sediment transport; and s* = the total erosion from all noncropland sources vector.

The sections of the model are described as follows. Equation (6.1) is the total cost function for agricultural production, including costs of production for crop and livestock commodities and the transportation of those commodities to the demanding consumers. Relation (6.2) is a resource transformation matrix relationship which converts the resource vector b, composed primarily of land and water resources, into the agricultural commodities. The commodities are provided to consumers through relation (6.3) from the final demand vector d. Relation (6.4) delivers residual soil loss as sediment from each producing area to the stream system. Sediment is then transported and collected in the stream system through relation (6.5). Relation (6.6) is the classical nonnegativity restraint for agricultural production. Finally, (6.7) and (6.8) represent environmental restrains on sediment delivery and transport to simulate environmental policy. Relations (6.1)-(6.3) and (6.6) are a part of the CARD-NWA model. Relations (6.4), (6.5), (6.7), and (6.8) are developed as a part of the sediment sector. The next section presents a detailed development of these relationships composing the sediment sector of the model.

SEDIMENT AND EROSION SECTOR

Three sources of erosion are included in this model. The erosion from cropland in production is the sum of the soil loss from cropping activities. Soil loss from cropland not in crop production is also estimated. Finally, erosion from all noncropland sources is fixed, since managing noncropland is not possible under this model.

Estimating soil loss for the production systems in each producing area follows the system explained for equation (1.2). The estimation of soil loss is determined by two variables which relate to cropland use and management. These are the C and P factors of the Universal Soil-Loss Equation (USLE). Other factors such as climate, soil type, and soil topography not altered by cropland management are fixed for this analysis at average annual values. The C factor includes two variables of cropland use: crop rotation system and crop tillage management practices. Up to 60 crop rotations per land quality group per producing area are available in the model for both dryland and irrigated agriculture. Rotations range from continuous production of corn, cotton, wheat, and other field crops to 5-year rotations of legume hays and field crops. The rotation and tillage practice determines the C factor for computing soil loss for each crop production activity.

Soil conservation practices (the P factor) are a second aspect of cropland management affecting soil loss. The four levels of conservation treatment defined in the model vary by producing area and land quality group. These are straight-row farming, contour farming, strip-crop farming, and terrace farming. The interaction of climate, based on the region of crop production, and crop rotations, tillage practices, and conservation practices provide a wide range of alternates for controlling sediment through cropland management.

Soil loss from land other than cropland is estimated for the year 2000. The total soil loss in each producing area from noncropland sources is computed by using a complete land use inventory (Nicol et al., 1974; USDA, 1971) and an extension of the Universal Soil-Loss Equation (USLE) (Wischmeier, 1965). The total soil loss from noncropland sources is fixed for the analysis and is expressed by the s* of equation (6.4). Details are provided in Wade and Heady (1976).

THE SEDIMENT TRANSPORT SYSTEM

The sediment water quality model includes three aspects of soil movement: sources, delivery, and transport. Sediment delivery and transport determine stream-level suspended sediment loads from the sources of soil loss. To approximate the movement of sediment, an interregional flow system for sediment links the producing areas of the agricultural sector. Since the producing areas are based on river sub-basins, the streams of each are assumed to flow uniformly into downstream producing areas or oceans.

Sediment Delivery

Sediment-delivery calculations account for the fact that only a small proportion of total soil loss from farmlands actually reaches streams. Equation (6.9) sums the soil losses from the cropland sources of erosion and the uncontrolled sources (uncontrolled in this model) and delivers the sediment to the river-basin system to be transported downstream. Thus the following relationship between sediment sources and sediment delivered exists for the ith producing area:

$$D_i\left(\sum_j \sum_k S_{ijk}X_{ijk} + S_i^* + \sum_j S_{ij}^A X_{ij}^A\right) = X_i^D \qquad (6.9)$$

where X_{ijk} = acres of production activity k on land quality group j; S_{ijk} = tons per acre of gross soil loss from activity k on land quality group j; s_i^* = tons of gross soil loss for land uses not endogenous to the model; X_{ij}^A = acres of idle cropland in land quality group j; S_{ij}^A = tons per acre of gross soil loss from the idle cropland of group j; D_i = proportion of gross soil loss that reaches the stream; and X_i^D = tons of suspended sediment delivered.

The entrapment of sediment in ditches, culverts, creeks, and small water-storage structures constitutes much of the "inefficiency" in the movement of sediment. The delivery ratio is fixed and determines the proportion of the gross erosion from all sources in each producing area that moves into downstream areas. The significance of such loads is determined by the stream geomorphology and the total downstream flow network.

Conventional procedures for computing delivery ratios proved inadequate for estimating sediment delivery ratios for the producing areas. (See Task Committee on Preparation of Sedimentation Manual, 1970, for typical methods.) Therefore, the delivery ratios for all producing areas are estimated by using data measured and computed for each producing area. This process brings many local variables and conditions into consideration in estimating delivery ratios. Table 6.1 gives the delivery ratios used for each producing area. Details of computation are given in a later section.

Table 6.1. Sediment delivery ratios used in the sediment transport submodel

Producing Area	Sediment Delivery Ratio	Producing Area	Sediment Delivery Ratio	Producing Area	Sediment Delivery Ratio
1	0.016	36	0.010	71	0.012
2	0.016	37	0.010	72	0.007
3	0.041	38	0.134	73	0.081
4	0.041	39	0.001	74	0.001
5	0.041	40	0.028	75	0.018
6	0.040	41	0.049	76	0.008
7	0.025	42	0.050	77	0.010
8	0.025	43	0.050	78	0.001
9	0.012	44	0.043	79	0.059
10	0.016	45	0.035	80	0.022
11	0.010	46	0.258	81	0.001
12	0.008	47	0.014	82	0.064
13	0.006	48	0.079	83	0.058
14	0.005	49	0.074	84	0.213
15	0.004	50	0.161	85	0.077
16	0.003	51	0.322	86	0.023
17	0.003	52	0.003	87	0.001
18	0.002	53	0.007	88	0.010
19	0.016	54	0.032	89	0.010
20	0.019	55	0.032	90	0.010
21	0.012	56	0.032	91	0.010
22	0.030	57	0.112	92	0.010
23	0.030	58	0.037	93	0.043
24	0.030	59	0.037	94	0.010
25	0.030	60	0.111	95	0.057
26	0.030	61	0.074	96	0.068
27	0.030	62	0.030	97	0.010
28	0.030	63	0.024	98	0.010
29	0.030	64	0.032	99	0.378
30	0.030	65	0.004	100	0.021
31	0.064	66	0.022	101	0.003
32	0.030	67	0.010	102	0.018
33	0.030	68	0.019	103	0.107
34	0.185	69	0.053	104	0.005
35	0.030	70	0.006	105	0.010

Sediment Transport

Once delivered sediment, X_i^D, is in the stream, it moves with the water in accordance with the physical properties of the stream. This characteristic is expressed by the following equation:

$$T_i\left(\sum_{l\mapsto i} X_l^D + \sum_{k\to i} X_k^T\right) = X_i^T \qquad (6.10)$$

where X_l^D, = sediment delivered from the lth upstream producing area directly into producing area i; X_k^T = sediment transported through an upstream producing area k into the stream system of producing area i; T_i = proportion of sediment moved to the boundary of producing area i that is transported through producing area j; and X_i^T = sediment transported through producing area i. The notation $l\mapsto i$ and $k\to i$ means "for all l (or k) contributing to i."

Sediment transport refers to the movement of sediment within the mainstream of a river system. In an open stream channel with no water impoundments or structures, most of the sediment load entering the stream moves with the water out of the region. Sediment transport is the proportion of all sediment delivered to a producing area from all upstream producing areas that leaves the producing area. All producing areas with such flow-through have transport ratios as shown in Table 6.2.

Table 6.2. Sediment transport ratios used in the sediment transport submodel

Producing Area	Sediment transport ratio	Producing Area	Sediment transport ratio
8	1.000[a]	59	1.000[a]
31	0.513	60	1.000[a]
34	0.735	63	0.270
38	0.001[b]	64	0.228
40	0.700	66	0.110
41	0.400[c]	68	0.067
42	0.540	69	1.000[a]
43	0.950	73	0.026
44	1.000[a]	75	0.003
45	1.000[a]	78	0.106
46	1.000[a]	79	0.188
48	1.000[a]	81	0.334
50	0.029	84	0.038
52	0.001	86	0.016
53	0.838	93	0.007[c]
55	1.000[a]	95	0.256
56	1.000[a]	96	1.000[a]
57	1.000[a]		

[a]No mainstream structures in this producing area.

[b]Minimum sediment transport set at 0.001

[c]Adjusted after personal telephone conversation with U.S. Army Corps of Engineers personnel.

River Basin Accounting

The total sediment load X_t at the point of river basin outflow is:

$$\hat{X}_{i'} = X_{i'}^T + X_{i'}^D \tag{6.11}$$

or the sum of sediment delivered from the last of the ith producing areas in the river basin and sediment transported through the ith producing areas. Sediment transport ratios for a sample of the producing areas are illustrated in Table 6.2.

On converting (6.9), (6.10), and (6.11) to the more familiar linear programming format these equations are:

$$\sum_j \sum_k S_{ijk} A_{ijk} + \sum_j S_{ij}^A X_{ij} - \frac{1}{D_1} X_i^D = -S_{vi}* \tag{6.12}$$

$$\sum_{k \to i} X_l^D + \sum_{k \to i} X_k^T - \frac{1}{T_i} X_i^T = 0 \tag{6.13}$$

$$X_{i'}^T + X_{i'}^D = X_{i'} \tag{6.14}$$

The variables D_i and T_i are constants depending on the physiographic and hydrologic makeup of the producing area. The quantity on the right of (6.13) is a constant depending on the geomorphology and land use of noncropland in the producing area. The delivery and transport of sediment have no direct cost in the model, either to individuals or to society as a whole. However, policies placing restrictions on sediment water quality will place implicit or implied costs on society by altering the production processes acceptable in satisfying the environmental goals of the policy.

METHOD OF SEDIMENT SECTOR PARAMETER ESTIMATION

The variables of the sediment transport subsystem are highly interdependent and cannot be estimated unless they are considered as a matrix of components. The estimated variables of the sediment transport system are based on data from 1960 to 1969. This period provides the best available land use data for computing total soil loss in the Conservation Needs Inventory (CIN) (USDA, 1971), which in turn gives the most comprehensive subdivision of land by quality and erosiveness of any currently available data.

Parameter estimation has three interrelated parts. First, for the base period, estimates are made of the total annual gross soil loss from all sources for each of the 105 producing areas. Second, sediment trapping in interregional flows of water is estimated. The final step is to compute the proportion of the sediment eroded from the lands of each producing area that enters the stream system from that producing area.

Total Gross Soil Loss

The most complex computations of parameter estimation involve estimates of sheet and rill erosion, gully erosion, channel erosion, and total gross erosion from all sources in each producing area. The process specifies an erosion source, determines its land-use and management characteristics, and computes the gross soil loss for the source. The total gross erosion is the sum from all sources.

Erosion originates from cropland and noncropland. This categorization divides the problem of erosion and agroenvironmental policy in the model into controllable and uncontrollable erosion. Land use data on the sources of erosion are of two types: inventoried and noninventoried. Inventoried lands are the privately owned lands surveyed extensively and classified according to several uses in the CNI. These lands are of seven specific types (cropland in rotation tillage [dryland and irrigated], other cropland, pasture lands, range lands, forest lands, grazed forest lands, and other lands) and are subdivided by specific land-use capability and conservation treatments needed. Less information is available for the noninventoried lands, which are classified according to three classifications: urban and builtup, federal noncropland, and small water areas.

All cropland sources of erosion in the system are based on inventoried cropland use, conservation treatment need, and land capability class. Cropland in each use subclassification and each land quality class is summed into two groups: treatment adequate and treatment not adequate. The total acres for each use subclassification are multiplied by an appropriate soil-loss rate to give the total soil loss, as specified in the USLE.

Erosion from the inventoried noncropland sources was determined by utilizing the land-use information from the CNI, SCS data, and an extension of the USLE. Soil-loss estimates are made for the noninventoried land uses by combining the land-use data of the national inventory with the soil-loss rates of similarly inventoried sources in the same producing area. Gully and channel erosion are computed as a fixed proportion of total sheet and rill erosion for each producing area.

Total gross erosion for each producing area is the total soil loss in an average year by all land uses described in the national inventory. This total gross erosion in the base historic period in the ith producing area is:

$$TS_i = CL_i + NC_i + NI_i + GE_i + CE_i \qquad (6.15)$$

where CL_i = soil loss from inventoried cropland sources; NC_i = soil loss from inventoried noncropland sources; NI_i = soil loss from noninventoried sources; GE_i = soil loss from gully erosion; CE_i = soil loss from channel erosion; and TS_i = total gross erosion. This equation is used to compute the sediment delivery ratios used in the sediment transport system of the model.

Sediment Transport Ratios

Sediment transport ratios are the aggregate transport efficiency of each producing area that has inflow from upstream producing areas and outflow to other producing areas or the ocean. A basic assumption used is that in the long run, sediment deposited in aggrading processes is offset by sediment removed in degrading processes.

U.S. Geological Survey Water-Supply Paper 1938 provides the locations of most of the major reservoirs and locks and dams in the United States (Martin and Hanson, 1966). The trapping efficiency of a reservoir is defined as the proportion of sediment flowing into a reservoir that is trapped (Brune, 1953). The trapping efficiency for the jth reservoir is assumed to be:

$$TE_j = (1 + 0.1 \cdot DU_j/DA_j)^{-1} \quad (6.16)$$

where TE_j = trapping efficiency; DU_j = storage capacity in acre feet of water; and DA_j = drainage area above the reservoir in square miles.

Structures for which data are inadequate to compute the trapping efficiency are assigned a trapping efficiency of 5 percent. An aggregation of trapping efficiencies that compounds the effect of each reservoir as sediment moves downstream determines the producing areas' sediment transport ratios as follows for the ith producing area:

$$T_i = \prod^{j \epsilon i} (1 - TE_j) \quad (6.17)$$

where TE_j = trap efficiency of the jth reservoir; and T_i = sediment transport ratio.

Sediment Delivery Ratios

The average annual total gross erosion from all lands and the average sediment load measured at the point of streamflow from each producing area are required to develop the sediment delivery ratios. Measured sediment loads are adjusted for the sediment transported through each producing area, since sediment delivery and transport are considered simultaneously.

The sediment delivery ratio for the ith producing area is computed as:

$$D_i = (S_i - \sum_{l \mapsto i} T_i S_{l \mapsto i})/TS_i \quad (6.18)$$

where D_i = delivery ratio; S_i = measured sediment load at the point of stream outflow; $S_{l \mapsto i}$ = measured sediment load at the point of stream outflow from the lth producing area flowing into i; TS_i = total gross soil loss; and T_i = sediment transport ratio. The stream sediment loads are measured from historic data for each producing area and the data sources used to obtain these values.

The ratios were tested prior to application in the complete linear programming model to verify the consistency of flows and the distribution of sediment loads under assumed conditions. This verification process gave a successful appraisal of the sediment transport and delivery system.

SOIL LOSS FROM NONCROPLAND SOURCES

In the policy analyses considered, total soil loss from noncropland sources to the sediment transport system is fixed. The total of this exogenous soil loss for the ith producing area in the year 2000 is:

$$S_i* = EC_i' + NC_i' + NI_i' + GE_i' + CE_i' \qquad (6.19)$$

where S_i* = total exogenous soil loss; EC_i' = exogenous cropland gross soil loss; NC_i' = inventoried noncropland exogenous gross soil loss; NI_i' = inventoried exogenous gross soil loss; GE_i' = exogenous gully erosion; and CE_i' = exogenous channel erosion.

The individual components of (6.19) are analogous to those computed in the calibration system. The differences between the total exogenous soil loss and the total soil loss used to compute the delivery ratios are changes in areas in each land use. The land base for computing exogenous erosion is adjusted to show the changes in land use between 1977 and the year 2000 and excludes the cropland uses which are determined within the model.

EROSION FROM UNUSED CROPLAND

Croplands not required for agricultural production are estimated to erode at the same rate as land in conservation use not adequately treated in the same land quality group in the same producing area. This soil-loss rate assumes that land not required for production is unused, unmanaged, and subject to higher-than-average soil-loss rates.

ALTERNATIVE FUTURES

The five alternatives considered are listed in Table 6.3. The Unrestricted Alternative serves to model the agricultural sector in the year 2000 if it is assumed that there is no change in the control policies for stream sediment loads. Thus it is used as a baseline for comparison with the other alternatives. A second alternative, the Minimum Sediment Alternative, minimizes the total national sediment load and the total cost of producing agricultural commodities in a two phase procedure. With this procedure, the total national sediment load is minimized first. This sediment load is then an added restraint in the model as the total cost of producing and transporting agricultural commodities is minimized. This alternative provides an extreme environmental restraint where maximum stream quality is obtained through cropland management while meeting prespecified consumer demands. A third alternative is used to examine stream sediment load changes and the economic impacts of limiting the soil loss from all crop production activities in each producing area to the level, established by soil conservation experts, that conserves the productive capacity of the soil, called the tolerance or t value. This allowable soil loss is between 2 and 5 tons per year depending on land quality group and region of the country. The fourth alternative is the Production Area (PA) Limit Alternative, which restrains sediment load contributions from agricultural lands for each producing area to 80 percent of the sediment loads from the Unrestricted Alternative. This alternative is used to examine some of the potential changes in location of crop production and in the

Table 6.3. Description and objective for each of the proposed alternatives for controlling agriculture's contribution to sediment loads in U.S. waterways

Alternative	Description	Objective
Unrestricted Alternative	Baseline alternative with no limits on sediment loads.	Minimize total production and transportation costs.
Minimum Sediment Alternative	Minimization of sediment load in nation's waterways.	Two phase minimization of sediment load and production and transportation costs.
t Limit Alternative	Soil erosion losses limited to conservation tolerance (t) limits of Soil Conservation Service.	Minimize total production and transportation costs.
PA Limit Alternative	Producing area (PA) sediment load contribution limited to 80 percent of the level in the Unrestricted Alternative.	Minimize total production and transportation costs.
River Basin Limit Alternative	River basin sediment load contribution limited to 80 percent of the level in the Unrestricted Alternative.[a]	Minimize total production and transportation costs.

[a]River basins in the model are aggregations of contiguous producing areas.

management of cropland that might be used to reduce sediment loads. Finally, a River Basin Limit Alternative is formulated, which restricts the agricultural contribution to the sediment loads of each river basin to 80 percent of the contribution from the Unrestricted Alternative.

RESULTS

Sediment loads at the point of stream outflow of the 18 major river basins are compared with average historic loads in Table 6.4 for the five alternatives. In general, the sediment loads for the Unrestricted Alternative for the year 2000 are comparable to the historic loads. The sediment loads of the Upper Mississippi and Missouri river basins (Figure 6.2) do differ substantially from the historic values in Table 6.4. Two distinct mechanisms produce this result. First, the use of reduced-tillage cropping practices in the Unrestricted Alternative in the Upper Mississippi river basin reduces the erosion substantially. This result differs somewhat from historic practice, but follows trend changes in practices experienced in recent years. Secondly, the relatively large amounts of unused cropland in the

Table 6.4. Sediment loads from actual data and the five sediment control alternatives

River Basins	Historic[a] Sediment Loads	Alternatives				
		Unrestrictive	Minimum	t limit	PA Limit	River Basin Limit
	(thousand tons)					
New England	1,760[b]	1,703	1,528	1,661	1,667	1,666
Mid-Atlantic	4,807[b]	4,355	3,790	4,086	4,223	4,223
South Atlantic-Gulf	12,916[b]	11,298	10,150	10,416	11,007	11,006
Great Lakes	NA[c]	6,895	4,833	5,991	6,433	6,433
Ohio	58,800[b]	51,535	43,029	47,065	49,602	49,602
Tennessee	18,400[b]	13,200	11,128	11,990	12,706	12,706
Upper Mississippi	181,000	121,593	69,213	90,710	109,757	107,403
Lower Mississippi	344,000[b]	309,641	226,332	274,462	291,240	291,240
Souris-Red-Rainy	1,770[b]	1,018	533	903	897	897
Missouri	20,380[b]	106,184	62,899	81,367	97,145	92,216
Arkansas-White-Red	32,186[b]	27,495	22,789	24,853	26,444	26,443
Texas-Gulf	18,542[b]	12,543	9,644	11,534	11,933	11,932
Rio Grande	275	1,206	1,138	1,249	1,189	1,190
Upper Colorado	32,655[b]	28,871	28,439	28,867	28,776	28,776
Lower Colorado	97[b]	2,119	2,082	2,107	2,111	2,111
Great Basin	NA[c]	5,342	5,091	5,323	5,288	5,289
Columbia-North Pacific	18,467[b]	15,468	11,594	15,393	14,657	14,658
California	42,717[b]	44,233	42,354	44,127	43,818	43,818
Total sediment outflow	432,435	415,024	318,297	376,451	393,671	393,665

[a]Historic sediment loads are the sums of available sediment outflows from each River Basin as computed from average historical data. Sources are detailed in Wade and Heady (1976) (see Figure 6.2).

[b]Used to compute sediment outflow to the oceans surrounding the United States.

[c]Not available.

Figure 6.2. The major river basins by county boundaries.

Missouri river basin under the Unrestricted Alternative erode at higher-than-average rates, producing more sediment than if the land were cropped. The unused cropland results from the relatively low agricultural demands in the model.

The sediment loads generated by the Unrestricted Alternative provide a baseline for comparison with the other alternatives as restraints are placed on sediment loads. For the Minimum Sediment Alternative, total sediment loads are reduced by 23.3 percent compared to the loads of the Unrestricted Alternative. This relatively small reduction in total sediment load following such an extreme environmental restriction is because 308.3 million tons of the total sediment load originates from noncropland sources that are not affected by changing the cropland use in the model. (The model being developed by CARD for the USDA's 1985 RCA Analysis includes range and forest sectors, as well as conventional cropland.) The agricultural portion of the sediment load in the Minimum Sediment Alternative is reduced by 90.6 percent from the level in the Unrestricted Alternative.

The impact of noncropland sources of erosion varies among the river basins. The largest proportion of total erosion from noncropland sources is in the Southwestern United States, where between 93.1 and 98.3 percent of the Unrestricted Alternative's total sediment loads are from noncropland. In contrast, in the Midwest a larger proportion of the sediment load is from cropland sources.

The t Limit Alternative, which restricts the model's cropping activities to those that produce no more than the allowable soil-loss level, results in sizable changes in average sediment load for most river basins. The waterways in the highly erosive areas of the Midwest and Southeast have sharply reduced sediment loads. The areas of the West are less restrained than the Midwest and Southeast since soil loss from cropland is much less. In fact, sediment loads increase in some areas of the West in the t Limit Alternative as cropping adjustments are made. The national sediment load decreases to 376.5 million tons per year, a 9.3 percent reduction from the level in the Unrestricted Alternative.

The PA Limit Alternative forces a 20 percent reduction in the sediment-load contribution of each producing area without a prior restricting of the activities defined in the model. This alternative results in a 5.1 percent reduction in national sediment load compared to the solution from the Unrestricted Alternative.

Total sediment load for the River Basin Limit Alternative is almost identical to that of the PA Limit Alternative: a 5.1 percent reduction in the national sediment load. However, since the restrictions on sediment loads in the model are imposed at the river-basin level rather than the producing-area level, there are major differences between the two alternatives in the use of the cropland and the cost of sediment load reductions.

Land Use

The key to the control of nonpoint sediment pollution from agriculture is the use of cropland. Land use determines soil erosion losses and, eventually, the sediment load in waterways. Changes in the levels of important cropland-use variables are presented in Table 6.5. Land use in the model under the Minimum Sediment Alternative is significantly changed from its use under the Unrestricted Alternative. There is a marked increase in the use of reduced tillage and terracing. There is also a change in the intensity of continuous rowcropping. This change requires more of the available cropland to be cultivated than is in the Unrestricted Alternative to produce the commodities demanded in the model. The additional acres are needed to offset the lower yields of the row crops shifted to less productive lands because of the less intensive rowcropping on the most productive land. The extreme environmental restriction of the Minimum Sediment Alternative also alters crop rotations and the location of production of crops. Specifically, significant increases in economic advantage are noted in the western river basins with the Minimum Sediment restraint.

Similar but less dramatic changes are required in land use for the t Limit Alternative. Relatively small changes in land use are noted between the Unrestricted Alternative and the PA and River Basin Limit alternatives. The changes in land use for the PA Limit Alternative reflect the imposition of restraints at levels that require change within each producing area without regard to changes that may be possible in other producing areas of the same river basin. In the River Basin Limit Alternative, intrabasin tradeoffs in output and production cost are utilized to meet required sediment reductions.

Table 6.5. Summary of national land use data for the alternatives

Item	Unrestricted	Mimimum Sediment		t Limit		PA Limit		River Basin Limit	
	million acres	million acres	%[a]	million acres	%[a]	million acres	%[a]	million acres	%[a]
Tillage Practice:									
Conventional tillage									
Residue removed	48.3	56.8	17.6	46.8	-3.1	49.5	2.5	48.7	0.8
Residue left	144.9	79.3	-45.3	135.0	-6.8	147.0	1.4	142.1	-1.9
Reduced tillage	115.6	226.6	96.0	128.9	11.5	119.2	3.1	117.5	1.6
Conservation Practice:									
Straight row	94.1	65.7	-30.2	63.4	-32.6	80.0	-15.0	81.9	-13.0
Contouring	112.4	126.9	12.9	111.0	-1.2	109.5	-2.6	107.4	-4.4
Strip cropping	68.7	5.8	-91.6	61.7	-10.2	81.9	19.2	83.4	21.4
Terracing	33.6	164.4	389.3	74.5	121.7	44.3	31.8	35.7	6.3
Crop Type:									
Row	187.1	183.4	-2.0	186.3	-0.4	185.1	-1.1	186.6	-0.3
Close-grown	58.2	77.7	33.5	60.1	3.2	62.8	7.9	57.9	-0.5
All hay	52.2	99.5	90.6	56.9	9.0	54.0	3.4	57.3	0.2
Summer fallow	11.3	2.3	-79.6	7.8	-1.0	13.9	23.0	11.5	1.8
Cropped Land	308.8	362.8	17.5	310.6	0.6	315.7	2.2	308.3	-0.2
Unused Cropland	55.6	1.6	-97.1	53.7	-3.4	48.6	-12.6	56.0	0.7

[a]Percent change from the Unrestricted Alternative.

Crop Production

The national mix of crops produced differs considerably between the Unrestricted and the Minimum Sediment alternatives. Although stable in total acreage, the mix of row crops grown changes under the Minimum Sediment Alternative as more corn and sorghum grain is grown for livestock feed and less silage is raised since silage production results in higher erosion losses. Production of hay and small grain also increases to offset the reduced production of silage for livestock feeding. This same set of substitutions exists to some extent in all of the sediment control alternatives.

Cropland Utilization

The crops grown and technologies used for growing them determine the sediment load of each area of the United States. These land uses are summarized nationally and regionally in this section. Emphasis is given to crop selection and the use of conservation and tillage practices in each quality class for the sediment control alternatives.

Minimum Sediment Alternative. The total cropped land increases greatly for the Minimum Sediment Alternative, 17.5 percent over the Unrestricted Alternative. (See Table 6.5.) Greater acreage is required to offset yield decreases caused by shifts to the less erosive crops and the use of soil conservation practices. Expanded land requirements are met using less productive lands. Hence, more acres are required.

Comparison of the Minimum Sediment Alternative with the Unrestricted Alternative shows large interregional shifts in crop production are required to meet the commodity demands. Major increases in row crops occur in the Souris-Red-Rainy, Missouri, Arkansas-White-Red, Great Basin, and California-South Pacific basins, while decreases greater than 25 percent occur in the New England, Great Lakes, Upper Mississippi, and Lower Colorado basins (Table 6.6). These changes for the Minimum Sediment Alternative result from shifts to hay and close-grown crops in the river basins with erosive soils. Generally, in the river basins of the West with less erosive soils more land is shifted to row crops. The decrease in row crops in the Lower Colorado basin results from an increased requirement for hay and small grains as livestock production expands there. In the Missouri basin, close-grown, hay, and row crops expand into the large areas of cropland not used for crops in the Unrestricted Alternative. In the Upper Mississippi river basin, the increase in close-grown crops accompanies a decrease in row crops.

The use of crop production technologies in the Minimum Sediment Alternative to reduce the level of sediment loads augments these regional and land-use changes. In the Minimum Sediment Alternative, reduced tillage and terracing are the production technologies most commonly used to reduce sediment loads. The results show that extreme changes toward reduced tillage are required to lower national sediment load by 25 percent under the Minimum Sediment Alternative. (See Table 6.5.)

Table 6.6. Cropland in row crops for the Unrestricted Alternative and percent change of row crop acreage in the other alternatives

River Basin	Unrestricted		Minimum Sediment	t Limit	PA Limit	River Basin Limit
	Alternatives					
	(000 acres)	(percent of total land cropped)	(percent change from unrestricted alternative)			
New England	159	35	-99.4	-52.2	-23.9	37.7
Mid Atlantic	3,191	41	-4.6	-9.5	-4.8	-6.5
South Atlantic-Gulf	15,309	86	1.8	-18.3	-1.2	-1.3
Great Lakes	15,862	82	-28.8	-1.8	-1.4	-0.5
Ohio	21,971	84	-26.3	-1.8	-0.5	-0.2
Tennessee	2,074	91	-5.6	-2.1	-1.9	-2.8
Upper Mississippi	51,677	91	-18.7	-5.7	-2.9	-4.7
Lower Mississippi	13,617	81	-45.9	-41.2	-3.5	1.5
Souris-Red-Rainy	6,140	36	122.9	78.0	-1.4	4.9
Missouri	24,929	35	33.8	27.5	4.5	-0.2
Arkansas-White-Red	15,902	44	46.0	1.6	1.2	-1.3
Texas-Gulf	12,419	70	-17.9	-1.0	-5.8	-0.1
Rio Grande	845	68	14.6	-28.4	6.5	-0.8
Upper Colorado	417	48	4.1	5.8	3.8	0.0
Lower Colorado	478	48	-65.3	19.9	-1.3	5.2
Great Basin	247	17	62.8	26.7	-30.0	-6.5
Columbia-North Pacific	523	5	8.2	-3.3	-15.3	0.4
California	1,323	31	98.6	-2.9	-6.3	-2.6
United States	187,085	61	-2.0	-1.0	-1.0	-1.0

Land use by quality group is shown in Table 6.7. The Minimum Sediment Alternative requires the use of almost all available cropland. This expanded land use results from shifting to conservation and tillage practices that reduce average crop yields. The average soil loss, even on the lower quality lands, is reduced by the application of these production technologies.

Table 6.7. National total cropland use by land quality group for the five sediment control alternatives

Land Quality Group[a]	Unrestricted	Minimum Sediment	t Limit	PA Limit	River Basin Limit
I	34,201	34,382	33,962	34,280	34,374
IIe	76,216	77,104	76,937	76,401	76,084
IIs, IIc, IIw	81,936	84,182	83,529	83,638	80,355
IIIe	59,312	67,587	61,920	59,996	61,322
IIIs, IIIc, IIIw	37,590	48,454	34,349	39,272	37,175
IVe	15,801	28,750	16,375	17,557	15,272
IVs, IVc, IVw	2,424	9,136	2,794	2,921	2,393
all V	20	873	20	82	21
all VI, VII, VIII	1,296	12,302	749	1,596	1,310
United States[b]	308,808	362,782	310,645	315,727	308,317

[a]The land quality groups are defined using the Soil Conservation Service land capability classification scheme.

[b]Totals may not add because of rounding.

t Limit Alternative. Some expansion in the total land area cultivated is required to meet the soil loss controls in the t Limit Alternative. (See Table 6.5.) The increase to 310.6 million acres is composed of a slight decrease in row-crop production and an increase in the acreages of close-grown and hay crops. Idled land is significantly different only in the Missouri river basin, where it is reduced by 7 million acres. Although the national requirements for row crops are relatively constant in the t Limit Alternative relative to the Unrestricted Alternative, the regional distribution of row crops changes considerably. (See Table 6.6.) Large increases in rowcropped lands occur in the Missouri, Souris-Red-Rainy, and Great Basin basins to compensate for changes in cropping methods. Small changes in conventional-tillage residue left and reduced-tillage acreage offset each other. Straight-row and strip-cropped acres are replaced by large increases in terracing. (See Table 6.5.)

PA Limit Alternative. The total use of cropland for this alternative is about 2.2 percent higher than in the Unrestricted Alternative at 315.7 million acres. (See Table 6.5.) The composition of crops changes from the Unrestricted Alternative, however. Total cropland use increases as close-grown crops, hays, and summer-fallow all increase. An increase in total acreage occurs as sediment restraints are placed on every producing area. More land must be cropped since the sediment restraints require a less intensive use of land. The largest increase is in the Missouri river basin, which has almost 5 more million acres in crops. Row crops occupy 59 percent of all the cropland, a 2 percent decrease from their area in the

Unrestricted Alternative. In addition, there are regional changes in row-crop production as decreases occur in the eastern and western river basins and increases occur in the middle of the country. (See Table 6.6.)

Table 6.7 shows that the increases in the total land use occur on most land quality groups. The largest increases, however, are on land groups IIs, IIc, IIw, IIIs, IIIc, and IIIw. These groups are potentially erosive, but with proper erosion control practices can be cropped without excessive soil losses.

River Basin Limit Alternative. Total land use for this alternative is only about 0.2 percent less than in the Unrestricted Alternative. Only slight changes occur in the composition of row crops and close-grown crops. (See Table 6.5.) Total summer-fallow acreage is slightly increased for the nation. Row crops increase significantly, by 37.7 percent, only in the New England river basin. (See Table 6.6.) In other areas, relatively small changes are required to meet the sediment restraint. Tillage practices also change little from the Unrestricted Alternative. However, a shift from straight-row to strip-cropping technologies is generally required. Terracing is not increased as in other alternatives. Hence, the total costs of crop production are less under the River Basin Limit Alternative than for other control alternatives. This is possible since the river basin sediment load restraint allows more flexibility in choosing the location and method of reducing sediment. The sediment load in some producing areas may be increased while it is decreased in others to meet stream sediment load restraints. This type of choice is not available at the intra-river-basin level for the PA Limit Alternative.

Cost. The tradeoff between the total production and transportation cost of demanded commodities and the total sediment load in the nation is illustrated in Table 6.8. The cost data show that as the level of sediment control (as measured by the total national outflow of sediment) increases, the total cost of controlling sediment increases sharply. In addition, on examining the cost data for the PA and River Basin Limit alternatives, it is obvious that the same level of national sediment load may be accomplished at numerous levels of national cost.

Table 6.8. **Total annual cost of producing and transporting commodities and sediment loads for the five sediment control alternatives**

Item	Alternatives				
	Unrestricted	Minimum Sediment	t Limit	PA Limit	River Basin Limit
Total cost (million dollars)	31,932	45,332	32,863	32,034	31,958
Increase in total cost (million dollars)	0.0	13,400	932	103	26
Index of total cost (percent of Unrestricted)	100.0	142.2	102.9	100.3	100.1
National sediment load (million tons)	415.0	318.3	376.4	393.7	393.7
Decrease in sediment load (million tons)	0.0	106.7	38.6	21.3	21.3
Index of total load (percent of Unrestricted)	100.0	76.7	90.7	94.9	94.9

SUMMARY

The alternatives considered simulate the attainment of a variety of environmental goals for the year 2000. Each goal is attained by some change in the makeup of the agricultural system. The most drastic changes occur when the goal is minimization of the sediment outflow of all U.S. river basins under fixed requirements for agricultural commodities. To accomplish this extreme environmental goal, large-scale changes in agricultural production are required. Stream sediment loads are drastically reduced under the assumed conditions of this alternative in most areas of the country, with the total national sediment load reduced 23.3 percent. Significantly, almost all lands that can be terraced are terraced, large acreages are cropped using contouring, and reduced tillage is common among the tillage practices. Such shifts require almost the entire available cropland base to be cropped to make up for production lost with the shift to less intensive row-cropping systems for soybeans and corn. The livestock feeding system is also changed drastically as both grains (particularly small grains) and hay are substituted for silage in the rations. Total cost of producing agricultural commodities is 42.2 percent higher than in the Unrestricted Alternative.

The t Limit Alternative limits annual soil loss from crop production activities to the level that will allow production on the land to continue indefinitely. These levels are established by the Soil Conservation Service. Although not as extreme as the Minimum Sediment Alternative, several significant agricultural changes occur. Regions of the country normally experiencing low soil loss are at a comparative economic advantage. Thus some of the rowcropping traditionally in high-erosion areas such as the Southeast is shifted to areas of the West and Southwest. Consequently, sediment loads in the Southeast are significantly reduced while loads in western areas either increase or fail to decline significantly. Although shifts in crop production and changes in land use are significant, the land area needed for commodity production does not increase much above that required in the Unrestricted Alternative. Total commodity production costs are increased 2.9 percent while total sediment load decreases 9.3 percent compared to the Unrestricted Alternative.

The PA Limit and the River Basin Limit alternatives are designed to evaluate the consequences of varying the level of the controls to reduce sediment loads. In the PA Limit Alternative the sediment contribution from the croplands from each producing area is restricted to 80 percent of the level found in the Unrestricted Alternative. Each producing area modifies the technologies used and crops produced just enough to meet the locally required reduction in agriculturally produced sediment loads. The cost of producing all commodities increases only 0.3 percent and total land used for crops is only slightly higher than for the Unrestricted Alternative.

The River Basin Limit Alternative calls for a reduction in sediment load equal to that of the Unrestricted Alternative except that the 80 percent limit is placed on each major river basin. Although the reduction in total sediment load in the nation's waterways is identical to the PA Limit Alternative, the total cost of commodity production increases only 0.1 percent. Total land required for crop production actually decreases compared to that needed in the Unrestricted Alternative. This alternative points up one of the tradeoffs in achieving desired stream quality goals when various levels of policy administration are applied. Given that the national sediment load must be reduced and the flexibility to select the reduction each producing area within the river basins will have to bear, the model uses some producing areas as sediment control regions. The sediment control regions produce the less erosive crops, and their contribution to stream sediment loads is relatively greater than that of other producing areas. The national water quality goal is accomplished, but the cost is less than in the PA Limit Alternative.

Under the modeled linkage between cropland agriculture and stream sediment water quality, changes in agricultural practices can be used to reduce sediment loads and meet environmental goals. The cost to agriculture, and thus to society, can be high for extreme levels of environmental control. However, lesser control of agriculture as a nonpoint source of pollution can be obtained at a relatively small cost through the reorganization of crop production patterns among and within various agricultural regions and through changes in land management practices, particularly if the control policies are designed to account for regional comparative advantages across the U.S. agricultural sector.

REFERENCES

Browning, G. 1967. Agricultural Pollution–Sources and Control. In *Water Pollution Control and Abatement,* edited by T. I. W. Hines, Iowa State University Press, Ames.

Meister, A. D., and K. J. Nicol. 1975. A Documentation of the National Water Assessment Model of Regional Agricultural Projections, Land and Water Use and Environmental Interaction. CARD Special Report. Center for Agricultural and Rural Development, Iowa State University, Ames.

Nicol, Kenneth J. 1964. A Modeling Approach to the Economics and Regional Impacts of Sediment Loss Control. Unpublished Ph.D. dissertation, Iowa State University, Ames.

Nicol, Kenneth J., Earl O. Heady, and Howard Madsen. 1974. Models of Soil Loss, Land and Water Use, Spatial Agricultural Structure, and the Environment. CARD Report 49T. Center for Agricultural and Rural Development, Iowa State University, Ames.

Task Committee on Preparation of Sedimentation Manual. Committee on Sedimentation of the Hydraulics Division. 1970. Chapter IV. Sediment Sources and Sediment Yields. *Journal of the Hydraulic Division of the American Society of Civil Engineers*, 96 (HY6): 1283-1329.

U.S. Department of Agriculture, Conservation Needs Inventory Committee. 1971. National Inventory of Soil and Water Conservation Needs, 1967. Statistical Bulletin 461. Washington, D.C.

U.S. Water Resources Council. 1974. 1972 OBERS Projections, Vols. 1-7. U.S. Government Printing Office, Washington, D.C.

Wade, J. C. and E. O. Heady. 1976. A National Model of Sediment Water Quality and Agricultural Production. CARD Report. Center for Agricultural and Rural Development, Iowa State University, Ames.

Wischmeier, W. H., and D. D. Smith. 1965. Predicting Rainfall-Erosion Losses from Cropland East of the Rocky Mountains. USDA Agriculture Handbook 282. Washington, D.C.

CHAPTER 7. A STUDY OF SEDIMENT-CONTROL POLICIES FOR U.S. AGRICULTURE UNDER LOW AND HIGH EXPORT LEVELS

by Joseph C. Campbell and Earl O. Heady

THE ISSUES related to water quality and the level of agricultural exports arise, in part, because higher export levels require a more intensive and extensive use of the nation's land base. As additional land is put into production or land is cropped more intensively it is expected that soil erosion losses will increase. These higher erosion losses then result in heavier sediment loads in the nation's waterways, which, in turn, impair water quality. Projections for increased exports in the future and the desire for improved water quality in the nation's waterways have created a need for analysis of policies to reduce sediment loads attributable to the agricultural sector.

This study focuses on the differential impacts of reducing sediment loads in waterways by a regulatory measure versus a taxing scheme (Campbell and Heady, 1979). Alternative policies, including subsidies and cost sharing, also have been analyzed by Daines and Heady (1981). The conceptual framework of the study is that the application of a per unit tax on sediment entering a waterway from agriculture will result in a smaller aggregate cost of a specified decrease in total sediment loads than a regulatory policy that limits soil erosion losses equally across the entire agricultural sector. The lower cost with the tax scheme occurs because the marginal cost of reducing sediment loads is equalized across the whole agricultural sector. The regulatory policy will not have this result; thus, higher costs are expected to achieve an equivalent decrease in sediment loads.

The regulatory policy restricts crop production activities in the model to only those activities with annual gross soil losses of less than 5 tons per acre while the tax policy sets a tax on sediment actually reaching the main river basins so as to generate the same national sediment load as the 5 ton per acre regulatory policy. This analysis of the resource allocations in the agricultural sector under the alternative policy approaches is carried out under two export levels to measure the sensitivity of the results.

THE MODEL

The interregional competition model employed in this analysis is defined with a set of regions consistent with the land and water resources, the crop and livestock production possibilities, and the interregional interaction of U.S. agriculture expected in the year 1985. The basic model of the agricultural sector is that described in Chapter 3 of this book. In the specification of the model, the land resources of U.S. agriculture are defined in 5 land quality groups for each of the 105 producing areas representing homogeneous production conditions (Figure 3.1). Crop production activities are defined by these 5 land quality groups in each producing area. Contiguous producing areas, aggregated to form 28 market regions (in Figure 3.2), serve as the basis for the transportation submodel, which, along with regional demand sets, causes interregional interdependency and competition in the model. Livestock production activities for swine, dairy, beef feeding, and beef cows are defined in each of these market regions. All other livestock, including poultry, are exogenous to the model. All crop and livestock production activities are based on projected 1985 technologies. Fixed commodity requirements are defined by market region and are based on projections of population, per capita consumption, and exports for 1985. For this study, results are reported by the 7 regions shown in Figure 7.1.

Endogenous production activities of the programming model include barley, corn, cotton, legume hay, oats, sorghum, wheat, and soybeans in rotational combinations. (The exogenous crops are the same as those listed in Chapter 3.) The soil-erosion coefficient for each cropping activity indicates gross soil loss per acre per year under the tillage practices and crop combinations represented by that

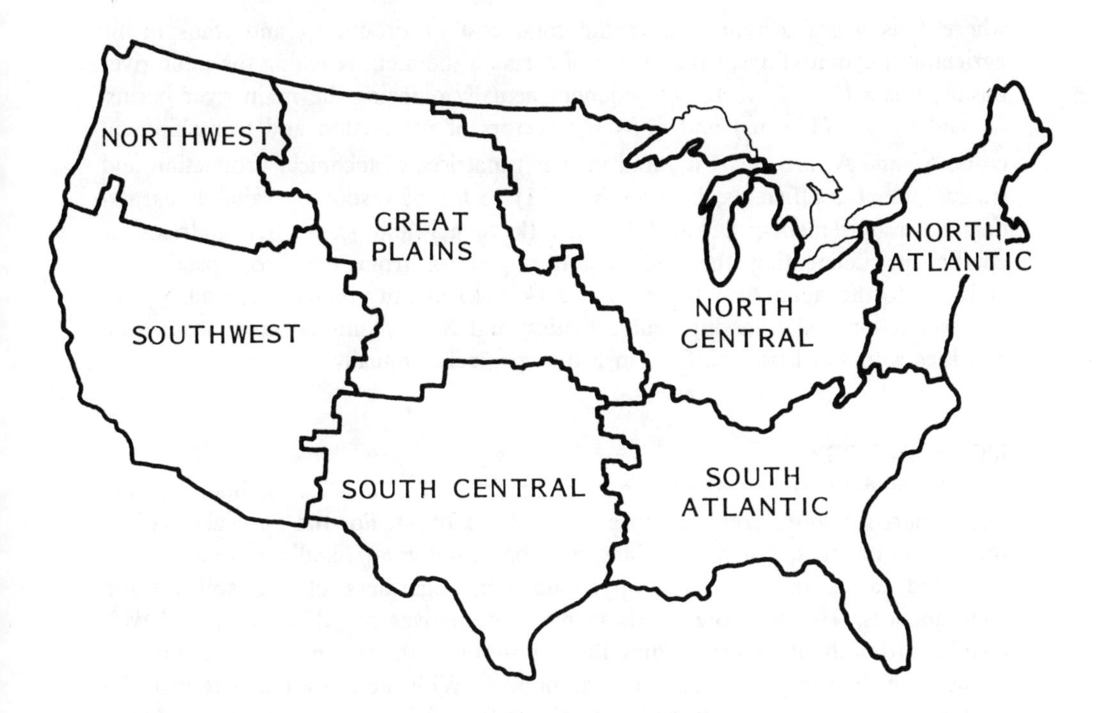

Figure 7.1. The seven reporting regions.

activity. As with the models discussed previously, production costs in these endogenous activities include market rates of return to all resources used in agriculture except land and water. Land and water returns are determined endogenously in the model.

The model includes the sediment delivery system developed in Chapter 6 and a set of activities associated with the sediment delivery system that allows a per unit tax on sediment to be incorporated into the objective function of the model. The resulting objective function thus minimizes the sum of both the cost of producing and transporting agricultural commodities and the tax charges on the total quantity of sediment delivered to the waterways from the crop production sector of the model. In matrix notation, the model can be written as follows:

$$\text{Minimize} \quad P + t \cdot S \tag{7.1}$$

$$\text{subject to} \quad (C_1C_2)(X_1X_2)' - 1P = 0 \tag{7.2}$$

$$(A_1A_2)(X_1X_2)' \geq b \tag{7.3}$$

$$(D_1D_2)(X_1X_1)' - IS = 0 \tag{7.4}$$

$$P, S, X_1, X_2 \geq 0 \tag{7.5}$$

where P is a scalar representing the total cost of producing and transporting agricultural commodities; t is a vector of costs on sediment reaching the main river basins; S is a (k × 1) vector of sediment actually reaching the main river basins; C_1 and C_2 are $(1 \times n_1)$ and $(1 \times n_2)$ vectors of production and transshipment costs; A_1 and A_2 are $(m \times n_1)$ and $m \times n_2)$ matrices of technical production and transshipment coefficients; b is an (m × 1) vector of resource availabilities and demand requirements; D_1 and D_2 are $(k \times n_1)$ and $(k \times n_2)$ matrices of coefficients delineating the sedimentation process from the crop production activities to the main river basins; I is a (k × k) identity matrix; X_1 and X_2 are the production and transshipment activities; and X_2 contains all crop production activities with soil loss greater than 5 tons per acre annually.

METHODOLOGY

The model is solved simulating no sediment-control policies for both low and high export situations (referred to as base alternatives). For the base alternatives there is no tax on sediments reaching river basin waterways, and agriculture is not restricted in its options for crop production, regardless of the soil erosion consequences. The solutions of these base alternatives are then compared with solutions from the model simulating the imposition of the sediment control policies to measure the various impacts of these policies. With the 5-ton per acre erosion-control policy all crop production activities with soil loss greater than 5 tons per acre annually are eliminated from the crop sector before the model is solved. The result is solutions in which the total costs of producing and transporting agricultural products have been minimized when soil loss can not exceed 5 tons per acre annually. In the sediment tax alternatives the tax in the model is set equal to $50. Thus the sediment tax alternatives minimize the total of the production and transportation costs and the taxes on sediment reaching the main waterways in the river basins. The $50 value was selected because this rate generates a total sediment load in the nation's waterways approximately equal to the total sediment load generated by the 5-ton limit alternative.

The export component of the demand sector is varied to investigate the sensitivity of the two sediment-control policies. The two export demand levels used in the study are shown in Table 7.1. The principal difference between the two export levels is a 25 percent change in the requirements for soybeans.

Table 7.1. Projected 1985 net exports used in the study

Commodity	Unit[a]	Export Quantities: Low export demand	Export Quantities: High export demand
		(millions)	
Corn	bushel	1,889	1,875
Sorghum	bushel	270	288
Barley	bushel	25	20
Oats	bushel	19	10
Wheat	bushel	1,179	1,137
Soybeans	bushel	1,125	1,397
Cotton	bale	4.2	4.1
Beef & Veal (carc. wt).	pounds	-1,190	-2,169
Milk (fresh equiv.)	pounds	-680	-680
Pork (carc. wt.)	pounds	-307	-307
Lamb & Mutton (carc. wt.)	pounds	-230	-230
Turkeys (R.T.C.)	pounds	70	70
Broilers (R.T.C.)	pounds	235	235
Eggs	dozen	43.9	44

[a]Carc. wt is carcass weight; fresh equiv. is fresh equivalent; and R.T.C. is ready-to-cook.

RESULTS

With the low export requirements both sediment-control policies greatly reduce total sediment load in the nation's waterways at a relatively small cost to consumers (Table 7.2). (The consumer cost calculation in Table 7.2 is a per capita value based on production and transportation costs from the model and does not include changes for processing, marketing, etc.) Reducing sediment loads under the high export requirements is more costly to consumers, especially with the 5-ton limit regulatory policy. This difference in consumer cost is the result of the resource allocation consequences of the two policies. The regulatory policy restricts the use of the land regardless of the value of the production foregone per unit of reduced sediment load. The tax policy approach incorporates this economic opportunity cost in the determination of regional resource allocations. Thus, if the value of a crop is sufficiently profitable in a region highly subject to erosion, the tax policy approach will result in soil erosion rates in excess of those allowed by regulatory policy. Given that the sediment tax rate is set to generate approximately the same total sediment load in the nation's waterways, the higher rates of erosion in some regions are offset with lower rates in other regions. The regulatory policy will produce a pattern of production that is more costly to the extent that there are differences in the cost of output foregone per unit of reduced sediment load in the model. The regional impacts of the two policies will be covered in more detail in later sections.

Table 7.2 also presents aggregate data on other variables for the agricultural sector from the various alternatives. Stressing the capacity of American agriculture with higher exports (the land resources are almost exhausted by the high export requirement) produces significantly higher per capita food costs. The higher prices imputed to land resources under the regulatory policy by the model also reflect the impact of the relatively inefficient economic result of the regulatory policy compared to the tax policy. The increases are especially dramatic under the higher exports.

Imposing the soil erosion control policies alters the national mix of crops grown. For example, fewer acres of silage and more acres of hay are grown as hay is substituted for silage in the livestock rations. This substitution occurs because of the relatively poor soil-erosion protection afforded the land by silage crops.

Land management and tillage practices also change under the sediment control policies (Tables 7.3 and 7.4). Straight-row farming and strip cropping decline relatively more under the regulatory policy than the tax policy. The use of contouring and terracing increases under both erosion-control policies, but relatively more under the regulatory policy than the tax policy. Generally, the regulatory approach results in the use of more expensive measures. In addition, the changes in both the crop mix and the land management practices required under the regulatory policy result in a greater use of fertilizer and pesticides than with the tax policy.

Table 7.2. Changes in major variables at the national level for all solutions (indexed to the Low Export Base)

	Low Export Alternatives			High Export Alternatives		
Item	Base	Tax Policy	Regulatory Policy	Base	Tax Policy	Regulatory Poli
Sediment Load	100	35	37	114	42	40
Per Capita Food Cost[a]	100	104	105	116	118	190
Land Cropped[b]	100	98	101	107	106	110
Land Shadow Price[c]	100	112	127	204	209	796
Silages[d]	100	84	76	98	81	66
Hays	100	110	116	103	115	122
Fertilizer Use	100	102	108	104	106	121
Pesticide Expenditures	100	123	141	118	152	201

[a]Calculation based on commodity prices and quantities generated by the model.

[b]Endogenous crops only.

[c]Dryland shadow price.

[d]Includes both corn and sorghum silage.

Table 7.3. The percentage distribution of land management practices for all solutions

Conservation Practice	Low Export Alternatives Base	Low Export Tax Policy	Low Export Regulatory Policy	High Export Alternatives Base	High Export Tax Policy	High Export Regulatory Policy
	(percent)					
Straight row	34.6	28.6	25.2	33.5	26.6	24.4
Contouring	49.4	51.0	52.8	46.4	46.4	50.6
Strip cropping	8.6	5.6	5.8	9.0	5.6	4.6
Terracing	7.4	14.8	16.2	11.1	19.3	20.4

Table 7.4. The percentage distribution of tillage practices for all solutions

Tillage Practice	Low Export Alternatives Base	Low Export Tax Policy	Low Export Regulatory Policy	High Export Alternatives Base	High Export Tax Policy	High Export Regulatory Policy
	(percent)					
Con RR[a]	17.1	15.1	15.6	17.5	14.9	17.6
Con RL[b]	48.8	44.6	38.2	47.8	44.8	33.9
Minimum tillage	34.1	40.3	46.2	34.7	40.3	48.5

[a]Conventional tillage with residue removed.

[b]Conventional tillage with residue left.

Regional and National Sediment Loads

The key environmental variable considered in this study is the sediment in the nation's waterways that can be attributed to production practices employed in the agricultural sector. The projected total sediment load from U.S. agriculture under the low export base is 83 million tons (Table 7.5). Of this total, 78 percent is in waterways in the South Atlantic region and 16 percent in the waterways of the North and South Central regions. The sediment load in the waterways of the South Atlantic region is especially heavy because the agricultural lands in this region are particularly subject to erosion and because the river flow system in the model carries a heavy sediment load from upstream regions. In the high export base the sediment load in the nation's waterways increases to 95 million tons. Relative to the low export base, the largest increase in sediment loads due to the higher exports is in the South Atlantic region, followed by the Great Plains and Western regions. These increases in sediment loads are related to the increase in total cropped land to meet the greater export requirement, the more intensive use of lands subject to erosion, and the fact that the principal difference between the low and high export alternatives is the 25 percent increase in the export requirements for soybeans, a very erosive crop.

The national sediment load in the 5-ton alternative is reduced by approximately 65 percent compared to the low export base. The $50 per ton tax on the sediment from agriculture in the sediment tax alternative produces a similar reduction in the national sediment load. Although the total sediment loads in the

Table 7.5. Regional and national sediment loads

Alternative	Low Export Alternatives		High Export Alternatives	
	Sediment Load	Change from the Base	Sediment Load	Change from the Base
	(thousand tons)	%	(thousand tons)	%
North Atlantic				
Base	790	100	824	100
Tax Policy	456	58	668	81
Regulatory Policy	330	42	326	40
South Atlantic				
Base	64,998	100	75,696	100
Tax Policy	20,731	32	25,177	33
Regulatory Policy	21,766	33	23,953	32
North Central				
Base	6,950	100	7,085	100
Tax Policy	4,434	64	5,015	71
Regulatory Policy	4,752	68	4,672	66
South Central				
Base	6,300	100	6,625	100
Tax Policy	2,257	36	2,619	40
Regulatory Policy	1,439	23	1,672	25
Great Plains				
Base	2,102	100	2,429	100
Tax Policy	446	21	781	32
Regulatory Policy	1,076	51	1,169	48
Northwest				
Base	1,527	100	1,762	100
Tax Policy	676	44	784	44
Regulatory Policy	849	56	971	55
Southwest				
Base	324	100	372	100
Tax Policy	249	77	239	64
Regulatory Policy	313	97	414	111
United States				
Base	82,991	100	94,793	100
Tax Policy	29,249	35	34,283	37
Regulatory Policy	30,325	37	33,177	35

nation's waterways are roughly equal under both policy alternatives, there are marked regional variations between the two policies. In the North Atlantic and the South Central regions the 5-ton limit policy reduces sediment loads more than the sediment-tax policy. In the Great Plains, Northwest, and Southwest regions the opposite is true. This differential impact of the policies is because the sediment tax, unlike the regulatory policy, controls pollution from various sources in accordance with the opportunity cost of output foregone per unit of reduced sediment loads. Within the interregional competition framework of the programming model, it is cost effective under the tax policy alternative that regional cropping patterns be such that relatively more erosion occurs in the North Atlantic and South Central regions than in the 5-ton limit alternative. Given that a tax rate was selected to generate approximately the same national level of sediment load as under regulatory policy, the heavier sediment load in these two regions is offset by reduced sedimentation in the Great Plains, Northwest, and Southwest regions.

Regional and National Cropping Patterns

The impacts of the sediment-control policies on national cropping patterns vary sharply with the level of exports. With low export requirements, the national acreage of row crops declines and small grains increases under both sediment-control policies, with the acreage of small grains increasing substantially more under the regulatory policy than the tax policy. The results obtained under the high export requirements differ sharply between policies. The total acreage of row crops declines and small grains increases under the sediment-tax policy, the same as in the low export alternative. However, with the regulatory policy the situation is reversed, as row-crop acreage increases while small-grain acreage declines (relative, of course, to the high export base). This result under the regulatory policy occurs because of substantial substitutions between feedstuffs in the endogenous livestock rations. Imposing the gross soil-erosion limit of 5 tons per acre uniformly across the agricultural sector substantially alters the relative cost of feedstuffs to the livestock sector causing the substitution in the rations.

In some regions there are substantial changes in cropping patterns (Table 7.6). In the South Atlantic region there is a substitution of small grains for row crops with both soil erosion control policies. The mix of crops grown in the North Central region changes only slightly except for a marked change in small-grain acreage under the regulatory policy. There are also changes in cropping patterns in the Great Plains region.

Table 7.6. Regional acreages of row crops and small grains obtained under each alternative

Region and Crop Type	Low Export Alternatives			High Export Alternatives		
	Base	Tax Policy	Regulatory Policy	Base	Tax Policy	Regulatory Policy
	(1,000 acres)					
North Atlantic:						
Row crops[a]	4,558	4,206	4,102	5,308	6,004	5,776
Small grains	5,098	5,125	5,241	4,571	3,656	2,814
South Atlantic:						
Row crops	36,344	30,600	25,370	39,662	33,591	27,490
Small grains	1,364	2,512	7,919	1,676	2,218	9,934
North Central:						
Row crops	101,937	102,661	99,064	109,597	108,426	114,547
Small grains	21,766	23,151	30,326	23,010	24,326	13,470
South Central:						
Row crops	32,883	42,541	24,714	33,206	29,933	31,928
Small grains	8,600	10,083	10,668	11,000	11,916	8,059
Great Plains:						
Row crops	22,250	19,762	33,835	33,848	30,755	40,764
Small grains	22,401	19,975	17,815	21,960	21,147	21,486
Northwest:						
Row crops	1,587	2,017	1,692	2,820	2,766	2,594
Small grains	6,902	7,415	6,240	7,602	7,230	6,745
Southwest:						
Row crops	2,171	2,189	2,159	2,254	2,236	1,280
Small grains	2,179	1,961	2,249	2,573	2,480	2,559
United States:						
Row crops	202,115	191,395	191,505	227,131	214,103	228,531
Small grains	68,319	69,330	80,466	72,415	72,980	65,077

[a]Includes corn and sorghum grain, silages, soybeans, and cotton.

Commodity Prices, Returns to Land, and Regional Incomes

The indices for the commodity shadow prices in Table 7.7 indicate the marked impact of the interaction between export levels and the soil loss control policies on the agricultural sector. With the lower export requirements, the sediment-control policies raise commodity supply prices less than 6 percent. Under the higher export level, the supply prices increase significantly, especially with the regulatory policy. The 5 ton per acre soil erosion limit raises the indices for commodity supply prices by about 60 percent more than the sediment-tax policy, even though total sediment loads are about the same. Because of the larger resource allocation resulting from the use of a regulatory policy, relatively more land is required to be cultivated to meet the given demands for agricultural commodities than with the tax policy. Thus supply prices are higher under the regulatory policy than the sediment-tax policy, especially with the higher export requirement.

Table 7.7. Indices of shadow prices for the endogenous commodities, and average food cost of endogenous crops per capita for all solutions (the low export base = 100)

		Low Export			High Export		
Item	Unit	Base	Tax Policy	5 ton limit	Base	Tax Policy	5 ton limit
Barley	Bu	100	107	107	130	140	274
Oats	Bu	100	107	119	144	137	278
Wheat	Bu	100	110	108	135	141	254
Corn grain	Bu	100	109	104	128	133	204
Sorghum grain	Bu	100	110	100	127	133	229
Oilmeals	Cwt	100	115	129	136	151	445
Cotton	Bale	100	104	109	115	116	184
Silages[a]	Ton	100	105	105	122	123	196
Legume hay	Ton	100	104	106	124	127	248
Nonlegume hay	Ton	100	106	107	128	131	248
Pork	Cwt	100	106	106	116	120	196
Milk	Cwt	100	103	102	108	110	141
Beef feeders	Head	100	103	104	116	116	185
Fed beef	Cwt	100	104	105	116	117	189
Nonfed beef	Cwt	100	104	105	116	117	189
Food cost per capita		100	104	105	116	118	190

[a]Silages include corn and sorghum.

The rent imputed to land in the model varies similarly as commodity supply prices. Rents imputed to land are the difference between commodity supply prices and the production cost of the cropping activity selected by the model by land quality group in each producing area. As shown in Table 7.8, there is considerable regional variation in the changes in land rents. The land-rent changes are smallest in the South Central region and largest in the Great Plains region. These differential changes are due to the regional differences in productivity and to the erosive characteristics of the lands in each region.

Changes in the idle land in each region under each alternative also reflect the regional differences in productivity and erosion hazard. Utilization of available

Table 7.8. Land rents (shadow prices) on dryland and percent uncropped land by region

Alternative	Low Export Alternatives		High Export Alternatives	
	Uncropped Land	Rent[a]	Uncropped land	Rent[a]
	%	dollars	%	dollars
North Atlantic				
Base	5.5	23.45	1.3	52.26
Tax Policy	5.5	32.35	3.0	54.07
Regulatory Policy	6.0	28.37	4.2	152.99
South Atlantic				
Base	9.7	21.60	3.6	47.89
Tax Policy	13.7	23.66	7.9	47.81
Regulatory Policy	12.9	23.92	5.4	174.71
North Central				
Base	5.9	25.48	3.0	53.68
Tax Policy	5.2	29.43	2.9	54.11
Regulatory Policy	4.4	35.23	1.8	212.14
South Central				
Base	10.6	17.12	2.6	30.93
Tax Policy	13.0	18.29	3.5	30.49
Regulatory Policy	10.0	18.23	1.6	74.81
Great Plains				
Base	18.7	10.94	7.4	23.92
Tax Policy	22.6	10.66	8.1	22.73
Regulatory Policy	14.4	14.88	4.5	114.88
Northwest				
Base	11.0	12.55	7.4	19.80
Tax Policy	13.9	11.53	12.8	21.59
Regulatory Policy	13.9	15.00	2.6	75.16
Southwest				
Base	14.3	5.52	7.7	7.43
Tax Policy	15.7	6.30	5.4	8.59
Regulatory Policy	12.9	8.02	2.9	46.28
United States				
Base	10.4	20.41	4.2	41.66
Tax Policy	12.1	22.85	5.4	42.58
Regulatory Policy	9.2	25.91	2.9	162.51

[a]Land rents (shadow prices) expressed in 1972 dollars.

land in the North Central region is very high relative to other regions. The average land rent for the North Central region is also the highest in the model. These two results provide a good measure of the high productivity of the cropland in this region. The North Central region has fewer idle acres under both policy alternatives compared to the base alternatives. Other regions show a different result. For example, in the South Atlantic region the total acreage of idle land increases under both policy alternatives–just the opposite of the North Central region. These results indicate the changes in comparative advantage of these two regions that occur with the imposition of national polices to reduce sediment loads in the nation's waterways.

An aggregate measure of the regional impact of the sediment-control policies is made by summing the value of the output of the endogenous commodities produced in each region. Table 7.9 presents the percentage each region produces

Table 7.9. The percentage each region produces of the total value under endogenous agricultural production[a]

Regions	Low Export Alternatives Base	Low Export Tax Policy	Low Export Regulatory Policy	High Export Alternatives Base	High Export Tax Policy	High Export Regulatory Policy
	(percent)					
North Atlantic	5.6	5.5	5.2	5.4	5.2	4.7
South Atlantic	11.1	11.6	11.7	11.1	11.4	10.7
North Central	29.5	31.2	29.7	29.1	30.8	33.7
South Central	25.1	24.6	25.4	25.3	25.0	24.0
Great Plains	18.6	17.5	18.2	19.4	18.1	16.4
Northwest	2.8	2.8	2.9	2.8	2.7	3.8
Southwest	7.3	7.2	7.0	6.9	6.8	6.7
Total	100	100	100	100	100	100

[a]Values are based on total production costs from the model's solutions; the product of quantities and supply prices.

of the total value of the endogenous agricultural production in the model. These regional percentages show only slight variations from alternative to alternative, except for the high export base. With high exports, the North Central region gains a larger proportion of the total value of U.S. agricultural production, while the Great Plains regions is slightly disadvantaged.

CONCLUSION

The results from the interregional competition model imply that significant expansion of exports can sharply raise commodity prices and increase returns to landowners. In addition, sediment loads in the nation's waterways can increase dramatically, thus affecting water quality. The imposition of sediment-control policies on the model's agricultural production sector indicate that changes in production practices and patterns have the potential to greatly reduce sediment loads from agriculture.

The differential impacts of the two policies chosen for analysis here provide a measure of the potential benefits in terms of less expensive food for consumers to be gained by adopting policies that tend toward an equalization of the marginal cost of reducing the sediment loads across the nation's waterways. These benefits are especially dramatic if the limits of the productive capacity of the agricultural sector are being approached.

REFERENCES

Campbell, J. C., and E. O. Heady. 1979. Potential Economic and Environmental Impacts of Alternative Sediment Control Policies. CARD Report No. 87. Center for Agricultural and Rural Development, Iowa State University, Ames.

Daines, David R., and Earl O. Heady. 1980. Potential Effects and Policy Alternatives on Regional and National Soil Loss. CARD Report No. 90. Center for Agricultural and Rural Development, Iowa State University, Ames.

SECTION IV. APPRAISAL OF THE LONG-TERM PRODUCTIVITY OF THE U.S. AGRICULTURAL LAND BASE

THE chapter in this section provides a review of two components of a study carried out as part of the 1980 appraisal under the Soil and Water Resources Conservation Act (RCA) of 1977. The study is made using an interregional competition linear programming model of the U.S. agricultural sector. The components of the appraisal presented include an evaluation of maintaining the productivity of the agricultural land base to the year 2030 through mandatory soil erosion control measures. A second component of the RCA study reported in this chapter is an assessment of the economic potential for the conversion of pasture and forestlands to cropland by the year 2030. The sensitivity of this long-run assessment is provided by varying the projections for the rate of improvement in crop yields.

CHAPTER 8. ANALYSIS OF LONG-TERM AGRICULTURAL RESOURCE USE AND PRODUCTIVITY CHANGE FOR U.S. AGRICULTURE

by Burton C. English and Earl O. Heady

TO PROTECT the quality and quantity of basic resources, Congress passed the Soil and Water Resources Conservation Act of 1977 (RCA) (Public Law 95-192). This act directs the Secretary of Agriculture to appraise continuously the condition of the nation's soil, water, and related resources and to develop a program to maintain or improve their condition. The appraisal requires continuous evaluation of the nation's resources and includes data on quality and quantity of soil, water, and related resources; data on the capability and limitations of those resources for meeting current and projected demands; data on current federal and state laws, policies, programs, rights, regulations, and ownership and trends of use, development, and conservation of these resources; data on changes, and the condition of these resources resulting from past uses; data on costs and benefits of alternative soil and water conservation policies; and data on alternative irrigation techniques regarding costs, benefits, impacts on soil and water conservation, crop production, and environmental factors. This appraisal is to be conducted in 1980 and every five years after this date. The data collected are to be used in analyzing, evaluating, identifying, and investigating the soil and water conservation programs. This chapter reviews components of a study carried out by the Center for Agricultural and Rural Development (CARD) as part of the 1980 appraisal of the RCA of 1977. The study is made using an interregional competition linear programming model of the U.S. agricultural sector.

STUDY'S OBJECTIVES

This 1980 study was made in conjunction with the Soil Conservation Service (SCS) and the Economic Research Service (ERS) of the U.S. Department of Agriculture (USDA) and the Resource Conservation Act's Coordinating Committee, with the objective of providing inputs into the evaluation process of Public Law 95-102 for the 1980 appraisal. A total of 69 alternative combinations of soil-loss control programs and levels of resource demands or commodity supplies were analyzed.

This chapter examines 7 of these alternatives. These selected alternatives include an analysis of the impact of allowing the development of potential croplands in conjunction with a range of projections for the rate of improvement in crop yields for the year 2030. Also included is an examination of the impacts on U.S. agriculture of several levels of allowable soil loss for the year 2030.

The remainder of this chapter provides an explanation of the model and data sources and the methodology used in developing the coefficients of the model. A mathematical explanation of the model's objective function and the producing area, market region, and national equations is also provided. The alternatives analyzed and the results obtained are discussed in the final section of the chapter.

THE MODEL, DATA SOURCES, AND METHODOLOGY USED IN RESTRAINT MODELING

The American agricultural sector varies widely in climate, soil types, farm practices, and farm structure. Due to these variations, a model designed to examine this sector should reflect the regional aspects of agriculture. The regions should be consistent with the characteristics needed to specify available resources, production techniques, and possible interregional interactions. Included in the model are 105 producing areas and 28 market regions illustrated in Figure 3.1 and Figure 3.2, respectively. Within these regions, constraints are defined for resource availability, commodity production, and commodity demands. The basic units of the programming model are the 105 producing areas, with 5 land groups each derived from the U.S. Water Resource Council's 99 aggregated subareas. Activities are defined by land groups (U.S. Water Resources Council, 1970). The 105 producing areas aggregate into 28 market regions. The market regions also serve as the market framework for the nitrogen-purchasing activities. Activities representing alternative production possibilities simulate crop rotations, soil conservation and tillage practices, water transfer and distribution, commodity transportation, and nitrogen supplies. In addition, producing areas 48-105 also define water supply regions.

A final set of regions is defined for reporting purposes by aggregating adjacent market regions into 8 major zones (Figure 8.1). The zones include the Northeast, Southeast, Lake States, Corn Belt, Delta States, Northern Plains, Southern Plains, Mountain, and Pacific. (In this chapter the Northern Plains and Mountain zones will be referred to as the Northern Plains.) These zones differ somewhat from those in earlier chapters.

Numerous data sets need to be developed to specify the resource restraints, the activities' coefficients and bounds, and the commodity demands. These various data sets are described in this section. For further details, the reader is referred to English et al. (1982).

The Land Restraints

The land base is the major resource constraint on the productive capacity of the agricultural system in the model. This resource base consists of three components: the number of acres of dryland and irrigated cropland readily

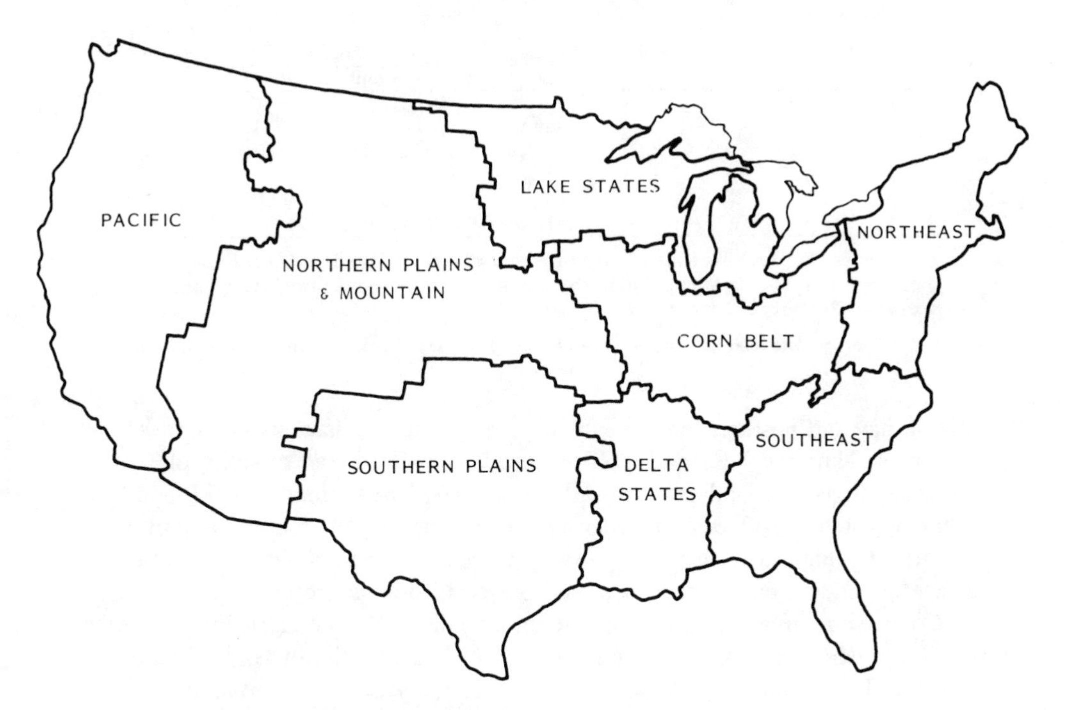

Figure 8.1. The eight major zones.

available, the number of acres that can be converted from dry to irrigated cropland, and the amount of land that can be converted from pasture and forestland to cropland. The last two components of the land base are incorporated in the model as activities and will be explained in a later section on coefficient and bounds development. This section will examine the development of the constraint equations defining the land that is readily available as cropland.

Both the Conservation Needs Inventory (CNI) (USDA, Conservation Needs Inventory Committee, 1971) and National Resource Inventory (NRI) (USDA, 1980) report the acres of privately owned land in the U.S. agricultural sector. These acreages are reported in both the CNI and NRI by the Soil Conservation Service (SCS) as eight major soil capability classes with classes II through VII further subdivided to reflect the most severe hazard preventing the land from being available for unrestricted use. The hazards reflect susceptibility to erosion, subsoil exposure, drainage problems, and climatic conditions preventing normal crop production. Thus, both the CNI and NRI report data by twenty-nine capability class-subclasses. The lands categorized by these twenty-nine capability class-subclasses are aggregated into 5 land groups as shown in Table 8.1. Only 5 land groups are used because of the magnitude of work to be done and the expected output of results.

The land base used for the endogenous crops represents the sum of the acres in the NRI defined as that used for row crops, close-grown crops, summer-fallow,

Table 8.1. Land classes and subclasses aggregated to five land groups

Land Groups	Inventory Class-Subclass[a]
1	I, II_{wa}[b], III_{wa}
2	Rest of II, III_c, III_w, III_s, IV_c, IV_s, IV_w, all of V
3	III_e
4	IV_e
5	all of VI, VII, and VIII

[a]Inventory classes and subclasses are as defined by the Soil Conservation Service for both the Conservation Needs Inventory and the National Resource Inventory.

[b]The wa indicates the wetness problem has been adequately treated.

rotation hay and pasture, temporarily idled cropland, and land used for fruits and vegetables. Thus the NRI and CNI are used to estimate the quantity of land in producing areas and land groups available in 1977. This estimated 1977 land base is then adjusted to reflect 2030 urban land needs, other nonagricultural land needs for airports, highways, vacation homes, recreation and wildlife, reservoir and surface mines, exogenous crop use, and projected double cropping.

Conversion of agricultural land to urban and other nonagricultural uses is considered when developing the cropland base. The quantity of land taken out of agricultural cropland to meet these demands in the year 2030 in millions of acres is 20.72, 0.67, 0.70, 4.64, 12.97, 1.93, and 6.27 (for a projected total for the United States of 47.9 million acres) for urban areas, highways, airports, second homes, recreational areas, reservoirs, and strip mines, respectively. These estimates are made by producing area and then disaggregated proportionately to the land groups within a producing area.

The endogenous crops grown in a producing area are those explained in Chapter 4. Crops such as fruits and vegetables are determined exogenously. For these crops, state irrigated and nonirrigated acreage projections from the National Interregional Projections (NIRAP) model were obtained, weighted to producing areas using the *1974 Census of Agriculture* (Bureau of the Census, 1977), and subtracted from the cropland base. These exogenous crops are categorized into three groups (close-grown, row crops, and orchard and vineyard) for both dry and irrigated lands. The projected exogenous crop areas are then distributed across the 10 land groups (5 for irrigated and 5 for dryland) by using the proportion of the three groups in each land group in each producing area.

The Water Restraints

The water restraints account for the predetermined supplies of water required for exogenous crops and livestock production. The projected irrigated acres producing exogenous crops provided by the NIRAP model are used in conjunction with water-use coefficients developed by the Special Projects Division (1976) to estimate the quantity of water required to produce the exogenous crops in the irrigated producing areas. (This represents a somewhat different set of water supplies than used in earlier models.)

The exogenous determination of livestock water demands is derived by using several sources. Livestock production was projected by state through the NIRAP model. These state projections are weighted from states to the producing areas with weights derived from the *1974 Census of Agriculture* (U.S. Department of Commerce, Bureau of the Census, 1977). Production by producing area is then multiplied by water consumption factors developed by the Agricultural Resource Assessment System Technical Committee (1975). These coefficients are then summed with the water required for irrigated exogenous crops to obtain the predetermined water requirements.

The Nitrogen Restraint

Exogenous crops require a significant amount of nitrogen. The nitrogen requirements for specific crops are multiplied by the exogenous crop acres in each market region to find the quantity of nitrogen required for exogenous crop production in each of the 28 market regions.

The procedure to determine the nitrogen fertilizer equivalent from livestock wastes accounts for the losses associated from handling, storage, and field application (Short and Dvoskin, 1977). These coefficients are estimated by market region for beef cows, beef feeders, dairy cows, pork, sheep, broilers, and layers. By subtracting the quantity of nitrogen supplied by livestock from the requirements of the exogenous crop production sector for nitrogen, the net contribution of the exogenous agricultural sector of the model is determined.

The Commodity Restraints

The final demand restraints, which are defined at the market region level, require the production of the endogenous commodities in the model to equal the projected levels of demand for the year 2030 (Table 8.2). The sum of domestic consumption, exogenous livestock feed demands, industrial and nonfood uses, and

Table 8.2. Projected 2030 demand levels used in the model

Commodity	Units	Total Quantity Demanded	Exports[a]
		---------(million units)------	
Barley	bushels	673.5	143.2
Corn Grain	bushels	9,437.5	3,077.1
Cotton	bales	12.3	3.3
Legume Hay	tons	103.8	0.0
Nonlegume Hay	tons	73.6	0.0
Oats	bushels	1,095.2	29.2
Oilmeals	cwt	2,012.9	871.2
Silage	tons	133.6	0.0
Sorghum Grain	bushels	1,490.6	424.4
Wheat	bushels	3,582.7	2,088.8

[a]These exports are included in the total quantity demanded.

exports determine these demand-restraint levels (English et al., 1982). The NIRAP model provides projections of domestic per capita consumption, livestock production, and net exports. The 2030 projections provided by the NIRAP model assume a domestic population of 300.3 million people (based on an annual rate of population growth of 2 percent). The projected population of 300.3 million people is distributed to the producing-area level based on a linear extrapolation of the projected population between 2020 and 2030. A population weight for each producing area is then calculated, and this weight, when combined with the national population projection, provides an estimated population by producing area. This population is then multiplied by the per capita consumption to obtain domestic food and fiber requirements by producing area. The producing area consumption figures are then summed to the market-region level.

National export figures (shown in Table 8.2) are used in conjunction with average port proportions to define the quantity of commodities exported by market region. The port proportions are determined from Consumer and Marketing Service (as referenced in Nicol and Heady, 1975). Importantly, the livestock sector is exogenous, not endogenous. Since livestock is exogenous to the model, the feedstuff demands are incorporated in the model using restraints at the market-region level. Concentrate and roughage rations are determined by using the 1971-73 state livestock production (Allen and Devers, 1975) and the 1971-73 average consumption of feedstuffs by livestock type (English et al., 1982). The concentrate rations determined from these two sources were compared to results from experimental feeding trials to project a 15 percent improvement in feed conversion for the year 2030. Roughage rations are determined similarly using roughage data provided by Allen. Since pasture is not endogenous, only hays and silages are included in the rations. Finally, as with the concentrate rations, an increase in conversion efficiency of 15 percent is assumed for the year 2030. Once the state livestock rations are determined, the state NIRAP livestock projections are used to calculate livestock feed demands.

The Crop Flexibility Restraints

Restraints are imposed on the model to limit shifts in production between producing areas. These restrictions on the model simulate farmers' multiple-cropping practices as a method of risk reduction and the imperfect mobility of resources. The adjustment restraints are applied to corn, cotton, sorghum, soybeans, and wheat. The adjustment limits are based on the crop production data reported in the *1974 Census of Agriculture* (U.S. Department of Commerce, Bureau of Census, 1977) and take the form:

$$.7 * QA_{iu}^{1974} \leq \sum_{j} \sum_{k} \sum_{m} X_{ijkm} W_{ijkmu} \quad (8.1)$$

i = 1, ..., 105 for the producing areas;
j = 1, ..., 10 for the land groups;

k = 1, ..., 330 for the rotation defined;
m = 1, ..., 12 for the conservation-tillage alternatives;
u = 2, 4, 11, 13, 15 for the crop.

where QA_{iu}^{1974} is the quantity of acres in producing area i planted in crop u in 1974; X_{ijkm} is the number of acres of rotation k with the conservation-tillage practice m in producing area i on land group j; and W_{ijkmu} is the weight of crop u in rotation k using conservation-tillage system m on land group j in producing area i.

THE COEFFICIENTS AND BOUNDS OF THE MODEL'S ACTIVITIES

Several sets of activities are used to define the endogenous crop-production sector for the model. The main set of activities in this sector are the cropping activities, which include rotations incorporating an objective function value, crop yields and soil loss, and operating inputs of labor, land, water, and nitrogen fertilizer. Additional activities in the crop-production sector include three types of water activities, a nitrogen purchase set of activities, two sets of land conversion activities, a set of irrigated hay conversion activities, and the transportation sector. These activities are described in more detail in the following sections.

Crop Production Coefficients and Soil-Loss Estimates

Unique crop production activities are defined on each land group in each producing area as explained in Chapters 1, 3, and 4. These activities define rotation systems of 1 to 4 crops, covering from 1 to 5 years, and employing selected conservation and tillage practices. The tillage practice options are fall plowing, leaving the crop residues until spring field preparation, or leaving the residue on the fields the year around (reduced tillage). The four conservation practice options are straight-row, strip cropping, contouring, and terracing. Thus for each rotation in each producing area there is a maximum of 12 crop-management strategies, each representing a unique combination of residue management and conservation measures.

Gross soil loss is determined by the Universal Soil-Loss Equation (USLE) as explained in equation 1.2 (Wischmeier and Smith, 1965). This soil loss does not represent the amount reaching the nation's waterways, nor does it incorporate that which is lost through wind erosion.

The KLSR values of equation 1.2 and percentage slope are weighted to the 5 land groups at the producing-area level, using the acres by land capability class and subclass and then aggregated to the land groups. The rotations and their corresponding factors are derived from the 1973 and 1978 SCS questionnaires as documented in English, Alt, and Heady (1982). The tillage C factors, conservation method, and the percentage slope(s) are then used to define the P factor. These are then combined to give a unique soil-loss coefficient for each rotation by conservation-tillage practice for each land group in each producing area. The soil-loss coefficient thus reflects the severity of erosion for the conditions appropriate for each defined cropping system.

The crop-yield coefficients. The crop sector in the RCA model is changed from previous models to directly account for the impact of cumulative soil-erosion losses on crop yields. The annual soil erosion for each cropping activity is projected using the Universal Soil-Loss Equation. Cumulative soil losses between the present and the year of projection (in this study 1980 to 2030) are then used to adjust projected crop yields downward with a procedure that is described in detail later in this section. For those activities selected in the model's optimal program of crop production activities for 2030, it is implied that these activities have been "in place" starting in 1980 by virtue of this relationship between cumulative soil erosion and crop yields in the year of projection.

The crop yield projection procedure involves several steps. The base yield for 2030 is determined using modified state Spillman functions developed by Stoecker (1974) and weighted to the producing-area level using the *1974 Census of Agriculture* (U.S. Department of Commerce, Bureau of Census, 1977). These yields are adjusted due to several criteria.

The 2030 state yields developed by Stoecker's Spillman functions take the form:

$$Y(t) = Y_o(t) + A(1 - .8^{X(t)}) \cdot PF(t) \qquad (8.2)$$

where A = the maximum potential yield response to fertilization; Y(t) = the estimated average yield per planted acre of the crop in year (t); $Y_o(t)$ = the estimated average yield per planted acre on unfertilized land in year (t) and developed from a linear trend function; X(t) = the number of units of fertilizer applied to each acre of the crop in year (t); PF(t) = the proportion of the acreage of the crop receiving fertilizer in the year (t) and developed from a linear trend of the proportion of the crop acres receiving fertilizer; and t is years after 1949. The X(t) above represents:

$$X(t) = Po(t) \cdot \{\ln (Px/Pc) - \ln A - [\ln(-\ln .8)]\} / \ln .8 \qquad (8.3)$$

where ln is the natural log of base e; Px is the weighted price of a unit of fertilizer; Pc is the price of a unit of crop c; and Po(t) is the proportion of the optimum rate of fertilizer applied in year (t) developed from a linear trend of the proportion of the optimum rates applied.

The above equations provide two components to the crop management system: an estimate of the optimal fertilizer application and the base yield. These data are then weighted to producing areas using weights developed from the *1974 Census of Agriculture* (U.S. Department of Commerce, Bureau of the Census, 1977).

Two adjustments are made to these yields depending on the rotation and the area of the country where the producing area is located. Carryover nitrogen from

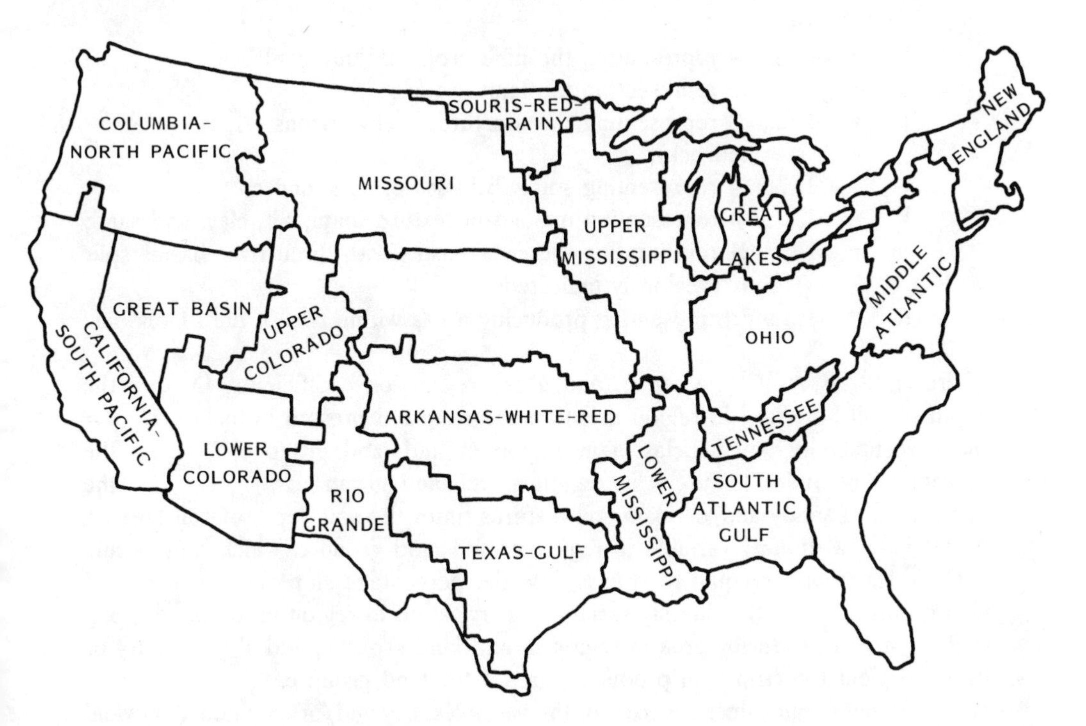

Figure 8.2. Yield adjustment regions.

legume sources is used in predicting the crop yields if legumes are defined within that rotation. The legume crop, especially legume hay, frequently produces more nitrogen than has been projected to be applied. When this happens, a new yield is determined, resulting in a higher yield than the projected base yield. The second adjustment to base yields occurs in summer-fallow regions. If a dryland rotation in a summer-fallow region has summer-fallow included in the cropping sequence, then yields are increased 5 percent. If the dryland rotation does not have summer-fallow, yields are reduced 19 percent.

The next step in determining crop yields is to adjust the yields for cumulative soil erosion losses between 1980 and 2030. To achieve this, Dyke and Hagen (1980) estimated dryland and irrigated yield equations for 21 regions of the United States (yield adjustment regions) (Figure 8.2).

The equations take the form:

$$Y_{clb} = \alpha_{mcb} + \sum_i \beta_{mcib} D_{mci} + \sum_j \gamma_{mcjb} L_{mcj} + \sum_k \theta_{mckb} T_{mck} + \mu_{mcb} V_{mc4} D_{mc1} + \delta_{mcb} D^{.5}_{mcl} + \epsilon_{mcb} D^2_{mc2} + \zeta_{mcb} S_{mc} + \eta_{mcb} S^2_{mc} + \lambda_{mcb} I_{mc} + \sum_n \alpha_{mcnb} P_{mcn} + \Sigma\, \varkappa_{mncb} P_{mcn} D^2_{mcn} \quad (8.4)$$

b = 1, ..., 9 representing the nine crops excluding silages;
c = 2, ..., 5 representing the land groups;
i = 1, ..., 3 representing the top three soil horizons (A, B, to bedrock);
j = 1, ..., 4 representing soil subclasses e, w, s, and c;
k = 1, ..., 4 representing major soil texture–loam, silt, clay, and sand;
m = 1, ..., 21 representing the river basins with three river basins split as previously indicated;
n = 1, ..., P representing producing areas within one of the 21 regions.

where α, ϵ, β, γ, ζ, θ, λ, η, κ, and μ are regression coefficients; D_{mci} = the depth of soil horizon i in region m and land group c expressed in inches; L_{mcj} = the percentage of soil subclass j in region m and land group c; T_{mck} = the percentage of major texture k in region n and land group c; $V_{mc4}\ D_{mc1}$ = the percentage of sandy and/or loam soil textures times the soil depth of soil horizon A; D^2_{mc2} = a dummy variable in region m and land group c, which only occurs if D_1 is less than or equal to two; S_{mc} = the percentage slope in region m and land group c; I_{mc} = the dummy variable for irrigation in region m and land group c; P_{mcn} = is a producing-area in region m and land group c; and Y_{clb} = a dry or irrigated yield for crop b in producing area 1 for land group c.

Using mean values for each of the variables, dry and/or irrigated 1980 yield for each producing area is determined for the crops defined in the model. The depth in soil horizon A is then varied to reflect the soil-loss rate of a given rotation. This is accomplished by taking the annual soil loss in tons, converting it to inches and subtracting the cumulative soil loss to the year 2030 from the mean soil depth. The yield resulting from this adjustment is divided by the mean yield and multiplied by the yield predicted by the Spillman function:

$$Y^u_{brnc} = Y^s_{brn} * \frac{Y^a_{bcn}}{Y_{bcn}} \tag{8.5}$$

where Y^u_{brnc} = the 2030 yield used for crop b in rotation r in producing area n and land group c; Y^s_{brn} = the 2030 yield predicted by the Spillman function for crop b in rotation r in producing area n; Y^a_{bcn} = the adjusted yield estimated from the regression function for crop b in producing area n and land group c; and Y_{bcn} = the mean yield estimated from the regression function for crop b in producing area n and land group c. Thus the yields in each rotation are adjusted for the productivity lost when soil is eroded. No function is developed for silages. When adjusting corn or sorghum silage yields, the equation for corn and sorghum is used.

Crop production costs. The data source for production costs is the Firm Enterprise Data System (FEDS) (USDA, Economic Research Service, 1976, 1977). These data are augmented with budgets from Agricultural Experiment Station offices and from Eyvindson (1965). The total costs are broken into four categories:

machinery, labor, pesticides, and other costs. Two adjustments are made for conservation and tillage practices. Timing factors are used to adjust machinery and labor costs for contouring, strip cropping, and terracing relative to straight-row cropping. Similar adjustments are made for tillage practices. The source used for developing these timing factors is the 1973 SCS questionnaire (Meister and Nicol, 1975). A second adjustment is made to reflect the tradeoff between tillage operations and pesticide use. When moisture is not deficient, it is assumed that a direct tradeoff exists between pesticide costs and the savings in machinery costs. In arid areas, the adjustment assumed consists of adding $3.00 in pesticide costs for each $1.00 decline in nonpesticide costs (Meister and Nicol, 1975).

The cost of terraces is incorporated in those cropping systems defined to include terracing. SCS data provide estimates of costs for terraces. The data are provided only for land groups 3 and 4 on which terracing cost is required. Costs are calculated using the two following equations:

$$TC_{ij} = .1(CC_{ij} + PW_{ij}W_{ij} + PT_{ij}T_{ij})PLT_{ij} \tag{8.6}$$

i = 1, ..., 105 for the producing areas;
j = 1, ..., 10 for the land groups.

where TC_{ij} = the per cultivated acre terracing costs on land j in producing area i; CC_{ij} = the per acre construction cost of terraces on land group j in producing area i; PW_{ij} = the proportion of acres of land group j terraced having grassed waterways for drainage in producing area i; W_{ij} = the cost per terraced acre for grassed waterways consistent with the terraces on land group j in producing area i; PT_{ij} = the proportion of acres of land group j terraced having tiled outlets for drainage in producing area i; T_{ij} = the cost per terraced acres of tiling and drainage consistent with terraces on land group j in producing area i; PLT_{ij} = the proportion of all land in group j which is feasible to terrace in producing area i; and .1 = the factor to adjust for a 10-year amortized life of the terrace.

$$C_{ijk} = \sum_m ((M_{ijm} + L_{ijm} + P_{ijm} + F_{ijm} + MS_{ijm}) *R_{ijm}) + TC_{jk} \tag{8.7}$$

i = 1, ..., the number of cropping management systems in the producing area;
j = 1, ..., 105 for the producing areas;
k = 1, ..., 10 for the land groups: 1, ..., 5 dryland and 6, ..., 10 irrigated;
m = 1, ..., 17 for only those crops in the cropping system.

where C_{ijk} = the cost per acre for crop management system i in producing area j on land group k; M_{ijm} = the projected per acre machinery cost for crop m in crop system i in producing area j; L_{ijm} = the projected per acre labor cost for crop m in cropping system i in producing area j; P_{ijm} = the projected per acre pesticide cost for crop m in cropping system i in producing area j; F_{ijm} = the projected per acre non-nitrogen fertilizer cost for crop m in cropping system i in producing area j; MS_{ijm} = the projected per acre other costs for crop m in cropping system i in producing area j; R_{ijm} = the rotation weight for crop m in cropping system i in producing area j; and TC_{jk} = the per cultivated acre terracing costs on land group k in producing area j (Meister and Nicol, 1975).

Fertilizer-use coefficients. The fertilizer coefficients used in the model are derived from solving the marginal conditions in the Spillman functions and are independent of land group, conservation practice, and tillage method.

Nitrogen Purchase Activities and Land Development and Conversion Activities

The set of nitrogen purchase activities in the model is unbounded to allow the crop sector to purchase as much nitrogen as required for the optimal solution. The level of fertilization required to attain the projected base yields is derived by subtracting the legume-provided nitrogen from that determined in the Spillman function. Nicol and Heady (1975) report the functions used in estimating nitrogen carryover. If a rotation has no legumes, then no adjustment is made to the nitrogen coefficient. If the rotation includes a legume, then the quantity of nitrogen provided by that crop is subtracted from the projected level of nitrogen needed.

The model incorporates several aspects of future land development and land conversion. Land development as discussed here includes both public and private development of land for irrigation. Land conversion includes the conversion of wetlands presently in pasture and forestlands to cropland.

Projected public irrigation developments in the model are 85 percent of the full-service acreage in authorized and funded projects expected to be in place by the year 2000. It is assumed that only replacement and maintenance funds will be available after the year 2000. Thus, the land base for 2030 assumes the same quantity of land development occurring from public investment as the year 2000 estimates used in studies reported in previous chapters. The projected areas of public irrigation are added to the available irrigated land base defined in the model. The quantity of land added is then subtracted from the quantity of land in the first four dryland groups. For additional information on these estimates see Meister and Nicol (1975).

Two types of land conversion activities are included in the model. These land conversion activities allow land development by producing area. One set of activities allows the model to determine if additional irrigation land is economical. A second set of activities converts forest and pasture wetlands, identified as being in the capability subclass II or III, to cropland.

Projected private irrigation development is not included in the irrigated land base. Instead, lands projected to be converted for irrigation by private development in the OBERS Series E' projections (U.S. Water Resources Council, 1975) are available for transfer to the irrigated cropland base through the economic decision processes of the linear programming model. The bounds for these activities are established by producing area consistent with the OBERS Series E' projections for the year 2020 (English et al., 1982).

The second set of land development activities in the model is for converting pasture and forestlands to cropland base on an average annual rate of conversion determined for each land resource area and the NRI acreages of Class IIw and IIIw pastureland and forestland in each producing area (English et al, 1982). This annual rate is applied to the declining balance of the wetland soils in projected acreage conversion by producing area to a maximum of 90 percent of the available acreage (Table 8.3). Conversion cost data for clearing and drainage are obtained from the 1973 SCS questionnaire and weighted by acreage conversion estimates to compute conversion costs by producing area (Meister and Nicol, 1975).

Water-use coefficients. The water-use coefficients for each irrigated crop management activity are estimates of the net diversion requirements for crop growth. Initially, gross diversion requirements were determined for each defined crop in each producing area by:

$$GDR_{iu} = \frac{CIR_{iu}}{(IE_u)(DE_i)} \qquad (8.8)$$

$i = 1, ..., 105$ for the producing areas;

$u = 1, ..., 17$ for the endogenous crops.

Table 8.3 Potential wetland development by river basin, 2030

River Basin	Maximum Pasture Conversion	Maximum Forest Conversion
	(thousand acres)	
New England	65.57	81.16
Mid-Atlantic	35.77[a]	481.17
South Atlantic	1,510.97	5,389.09
Great Lakes	663.20	1,374.18
Ohio	1,072.01	923.59
Tennessee	292.26	317.02
Upper Mississippi	1,739.30	1,315.51
Lower Mississippi	2,321.18	2,884.84
Souris-Red-Rainy	277.11	265.96
Missouri	1,092.88	234.82
Arkansas-White Red	405.31	242.06
Texas-Gulf	28.60	14.71
Rio Grande	0.00	0.00
Upper Colorado	0.00	0.00
Lower Colorado	0.00	0.00
Great Basin	0.00	0.00
Columbia-North Pacific	0.00	0.00
California-South Pacific	0.00	0.00
United States	9,504.16	13,524.11

[a]In some cases, exogenous land uses (i.e., urban expansion and exogenous crop use) required more land than the crop land base would permit. Thus, amounts of forest and pasture lands were reduced to meet these needs. Subsequently, the amount of lands available for conversion to cropland was reduced.

where CIR_{iu} = the quantity of water required by crop u in producing area i; IE_u = the irrigation or on-farm efficiency of crop u in using the water applied; and DE_i = the canal or delivery system efficiency between the diversion point and the farm. The gross diversion requirements are then used in determining the net diversion requirements by:

$$NDR_{iu} = CIR_{iu} + (1 - RF_i)(GDR_{iu} - CIR_{iu}) \tag{8.9}$$

$i = 1, ..., 105$ for the producing areas;
$u = 1, ..., 17$ for the endogenous crops.

where GDR_{iu} and CIR_{iu} have been previously defined and RF_i is the percentage of water not used by the crop and returned for reuse.

The Water-Sector Coefficients

Five sets of activities delineating the water sector are incorporated within the modelling framework. These activities include water purchase, water depletion, water transfer, water efficiency, and release of irrigated hay. The water purchase activities allow the buying of dependable supplies of surface and groundwater. The water depletion activities define the amount of groundwater available for depletion. Water that is left unused in a producing area can be transferred to downstream producing areas or through existing canals to other river basins with the water transfer activities. The water efficiency activities reflect the possibility of increasing the efficiency of water conveyance and application through investments amortized over a 10-year life. The release of irrigated hay activities allows irrigated exogenous hay land to revert to dryland hay. This conversion supplies additional water to the water sector and requires additional nonlegume hay to be produced on the endogenous cropland. For additional information on the water sector see Colette (1976).

The Transportation Sector

Interregional interdependence is obtained through the transportation sector. Transportation routes are defined among all contiguous market regions for barley, corn, oats, sorghum, soybeans, and wheat. In addition, long-haul routes are defined between nonadjacent market regions if mileage is reduced by 10 percent or more over the accumulated short-haul routes. Further information is available in Meister and Nicol (1975).

A MATHEMATICAL PRESENTATION OF THE MODEL

This section summarizes the mathematical relationships incorporated in the model. Model sectors already explained will not be illustrated. An explanation of the subscripts and the variables is provided at the end of this section.

The objective function minimizes the total cost of crop production, and transportation assumes competitive equilibrium as already explained. All resources receive market prices, except land and water, whose returns are determined endogenously. Costs in the objective function include labor, machinery, pesticides, fertilizers, water, and transportation of raw agricultural commodities. These costs are all specified in 1975 dollars. The objective function can be represented by:

$$\begin{aligned}\min OBJ = & \sum_i \sum_j \sum_k \sum_m X_{ijkm} XC_{ijkm} \\ & + \sum_r (W_r WC_r + W_r^E WC_r^E + IB_r IC_r + W_r^D WC_r \\ & + \sum_n \sum_s \sum_t T_{nst} TC_{nst} + \sum_i (LD_i DC_i + RD_i RC_i) \\ & + \sum_n F_n FC_n \end{aligned} \tag{8.10}$$

s = 1, 2, 8, 11, 13, 15 for the commodities transported;
t = 1..., 176 for the transportation routes defined.

Producing Area Equations

Each producing area has restraints for land availability separated into 5 dry and 5 irrigated land groups and restraints to control the level of production of 7 crops. In addition, water availability and soil loss are constrained at the producing-area level. The producing-area equations are as follows:

Dryland restraint by land group:

$$\sum_k \sum_m X_{ijkm} AD_{ijkm} - LD_{ij} + RD_{ij} \leq DA_{ij} \tag{8.11}$$

Irrigated land restraint by land group:

$$\sum_k \sum_m X_{ijm} AI_{ijkm} - RD_i RDP_{ij} \leq IA_{ij} \tag{8.12}$$

Crop acreage restraints:

$$MINA_{iu} \leq \sum_j \sum_k \sum_m X_{ijkm} W_{ijkmu} \tag{8.13}$$

In the producing areas in the western United States, water supplies and irrigation activities are defined. The following equation controls the allocation of water to the endogenously determined agricultural uses:

$$\sum_i \sum_k \sum_m \sum_u X_{rjkm} W_{rjkmu} CWU_{iu} + -WH_r WA_r - W_r^E I_r^E \leq WS_r \tag{8.14}$$

A final set of producing area constraints controls soil loss. The soil-loss equation is:

$$\sum_i \sum_k \sum_m X_{ijkm} SL_{ijkm} \leq ASL_i \tag{8.15}$$

Commodity Market Equations

To reflect demand based on per capita food use as a function of income, and commodity substitution and foreign trade movements through the region, each commodity market region has a set of equations to balance the supply and demand of the commodities. Additionally, a set of equations specifying the nitrogen fertilizer market is included by market region. The equations are:

Commodity balance equation:

$$\sum_i \sum_j \sum_k \sum_m X_{ijkmn} W_{ijkmu} CY_{ijkmsu} - \sum_t T_{nst} - \sum_r WH_r DA_{rs} \geq CD_{ns} \quad (8.16)$$

Nitrogen balance equation:

$$- \sum_i \sum_j \sum_k \sum_m X_{ijkmn} W_{ijkm} N_{iku} + 1 \geq NR_n \quad (8.17)$$

National Demand Equations

The national demand equations to be met are:

$$\sum_i \sum_j \sum_k \sum_m X_{ijkmu} CY_{ijkmu} \geq CD_u \quad (8.18)$$

i = 1, ..., 105 for the producing areas;
j = 1, ..., 10 for the land groups;
k = 1, ..., 330 for the rotations;
m = 1, ..., 12 for the conservation-tillage practices;
n = 1, ..., 28 for the market region;
r = 48, ..., 105 for the producing areas in irrigated regions;
s = 1, 2, 3, 5, 6, 8, 11, 12, 13, 15 for the commodities balanced at the market region;
t = 1, ..., 176 for the transportation activities defined;
u = 1, ..., 17 for the crops.

The endogenous commodities and their respective numbers used throughout the text are as follows: barley, 1; corn grain, 2; corn silage, 3; cotton, 4; legume hay, 5; nonlegume hay, 6; oats, 8; sorghum grain, 11; sorghum silage, 12; soybeans, 13; wheat, 15; and summer-fallow, 17.

The following are definitions of variables and indices mentioned in the text: AD_{ijkm} = the acres of dryland used per unit of rotation k using conservation-tillage method m on land group j in producing area i; AI_{ijkm} = the acres of irrigated land used per unit of rotation k using conservation-tillage method m on land group j in producing area i; ASL_i = the quantity of soil loss allowed in producing area i; CD_{ns} = the exogenously determined demand for commodity s in market region n; CWU_{iu} = the per acre water-use coefficient for crop u in producing area r; CY_{ijkmsu} = the per acre production of commodity s from crop u in rotation k using conservation-tillage system m on land group j in producing area i; DA_{ij} = the acres of dryland available on land group j in producing area i; DA_{rs} = the reduction in nonlegume hay yield associated with the conversion of an acre of irrigated pasture to dryland pasture in water supply region r; DC_i = the per acre cost for draining and converting land to cropland in producing area i; F_n = the number of pounds of nitrogen fertilizer purchased in market region n; FC_n = the cost per pound of nitrogen fertilizer purchased in market region n; IA_{ij} = the acres of irrigated land available on land group j in producing area i; IB_r = the acre feet of water transferred out of region r; IC_r = the cost differential on a per acre foot basis for water in region r; LD_i = the number of acres of land drained and converted to cropland in producing area i; LD_{ij} = the quantity of land drained in producing area i on land group j; $MINA_{iu}$ = the minimum acreage of crop u required in producing area i; RC_i = the cost per acre for private irrigation development in producing area i; RD_i = the number of acres developed for irrigation under private development in producing area i; RD_{ij} = the quantity of irrigated land developed in producing area i which is on land group j; SL_{ijkm} = the soil-loss coefficient for rotation k and conservation-tillage practice m on land group j in producing area i; T_{nst} = the number of units of commodity s transported over route t from market region n; TC_{nst} = the cost per unit of commodity s transported over route t from market region n; W_{ijkmu} = the weight of crop u in rotation k using conservation-tillage system m on land group j in producing area i; W_r = the number of acre feet of water purchased in water supply region r; W_r^e = the activity level of the increased irrigation efficiency in water supply region r; WC_r = the cost per acre foot of water purchased in water supply region r; WC_r^E = the cost of increasing the irrigation water efficiency in water supply region r; WH_r = the level of irrigated to dryland pasture conversion in water supply region r; WS_r = the acre feet of water available for use by the endogenous agricultural sector in water supply region r; X_{ijkm} = the number of acres of rotation k with the conservation-tillage practice m in producing area i on land group j; and XC_{ijkm} = the cost per acre of rotation k with conservation tillage practice m in producing area i on land group j.

ALTERNATIVES ANALYZED

As stated earlier, this chapter presents only a subset of the alternatives developed and analyzed under the 1980 CARD-RCA project. The first set of alternatives considered are part of an analysis restricting gross soil erosion at the producing-area level by 20, 40, and 60 percent of the levels found in an unconstrained alternative. The second analysis from the CARD-RCA project presented here allows lands classified in the National Resources Inventory (NRI) as of medium and high potential for conversion to agricultural production to be shifted to cropland. Included in this analysis are two alternatives in which the rate of improvement in crop yields is varied from the baseline projection to determine the sensitivity of the land conversion results. A listing of the several alternatives included in these two analyses is provided in Table 8.4.

Table 8.4. Listing of the alternatives and descriptions of the changes in the model formulations

Alternative	Change in Model Formulation
Base I	Unconstrained version of model.
20 Percent Reduction	Soil erosion losses constrained at producing area level to 80 percent of the erosion losses from solution of Base I.
40 Percent Reduction	Soil erosion losses constrained at producing area level to 60 percent of the erosion losses from solution of Base I.
60 Percent Reduction	Soil erosion losses constrained at producing area level to 40 percent of the erosion losses from solution of Base I.
Base II	Allows conversion of medium and high potential lands to agricultural production.
Low Crop Yields	Allows conversion of medium and high potential lands to agricultural production and assumes a lower rate of crop yield improvement than projected in Base I and Base II.
High Crop Yields	Allows conversion of medium and high potential lands to agricultural production and assumes a higher rate of crop yield improvement than projected in Base I and Base II.

In the soil erosion control analysis for the year 2030 the solution of the baseline alternative, Base I, provides both the starting point in constraining the alternative formulations of the model and the point of comparison for the comparative statics analysis of controlling soil erosion. The gross soil erosion in each of the producing areas is aggregated to 8 major zones from the solution to Base I and presented in Table 8.5. The maximum allowable erosion losses for the three alternatives restricting erosion losses are also presented in Table 8.5.

Table 8.5. Levels of allowed soil loss by major zone in 2030

Major Zone	Alternatives			
	Base I	20 Percent Reduction	40 Percent Reduction	60 Percent Reduction
	(million tons)			
Northeast	27.52	22.02	16.52	11.02
Southeast	54.85	43.88	32.91	21.94
Lake States	172.61	138.09	103.57	69.04
Corn Belt	138.01	110.41	82.81	55.21
Delta States	220.50	161.05	121.14	80.76
Northern Plains	264.59	212.98	160.59	107.51
Southern Plains	352.50	282.16	211.83	141.49
Pacific	52.48	41.19	32.36	23.54

The alternatives considered in the analysis of potential croplands identified in the NRI allow investigation of the comparative advantage of converting these forest and pasture lands to cropland use versus more intensive cropping of the existing cropland base. (See Table 8.6 for regional acreages of these NRI potential croplands.) The projected crop yields for the year 2030 are varied to provide a

Table 8.6. Aggregate values for variables of alternatives in the soil conservation analysis

Alternative	Endogenous Land Use	Row Crop Acres in Continuous Row Crops	Acres Fall Plowed and Cropped with Straight Rows	Expenditures[a]			Average Land Values[a]
				Fert.	Pest.	Total[b]	
	(million acres)	(thousand acres)	(million acres)	(million dollars)			(dollars per acre)
Base I	365.0	147,536	13.9	8,485	11,761	36,839	135
20 Percent Reduction	364.9	140,848	12.4	8,493	14,108	39,198	133
40 Percent Reduction	364.4	141,316	9.5	8,493	16,487	41,698	129
60 Percent Reduction	363.2	136,329	6.1	8,624	17,285	42,849	156

[a]1975 dollars.

[b]Total includes more than fertilizer and pesticides expenditures.

measure of the sensitivity of the results obtained in the land conversion analysis.

The adjustments in crop yields are based on the three trends incorporated in the estimating procedure described in the previous section: the unfertilized time trend for yields, the time trend for the percent of optimal fertilization of crops, and the time trend for percent of acres fertilized. For the lower rate of crop yield improvement these trends are adjusted downward to 90 percent of the trend between 1980 and 2030. For the higher rate of crop yield improvement these trends are adjusted upward to 160 percent of the trends between 1980 and 2030.

The next section of this chapter presents the results from the soil conservation analyses. The results from the land conversion analysis will be presented in the following section. The final section of this chapter is a brief discussion of the 1985 RCA study to be carried out by Earl O. Heady, Burton C. English, and other personnel of CARD.

RESULTS FROM THE 1980 RCA SOIL CONSERVATION ANALYSIS

Some aggregate results from applying the increasingly stringent restraints to reduce soil erosion are presented in Table 8.6. The total use of land for endogenous crops declines slightly as soil erosion is reduced by 60 percent from the Base I solution. Although the total acreage does not change much, the use of the land is altered markedly. For example, the practice of continuous rowcropping declines as the model is forced to reduce soil erosion. Other land management practices also change. The number of acres of cropland that are both fall-plowed and then cultivated using straight rows declines by more than 60 percent when the 40 percent alternative is compared to Base I. (Additional disaggregated data from the model results for land management practices will be presented later.) These changes in the way land is managed alter the aggregate mix of inputs employed in the crop production process. Total expenditures for fertilizer increase only slightly while expenditures for pesticides increase markedly.

The interaction between the various changes in land management practices and the use of marginal lands affects the value imputed by the model to land resources. As soil losses are reduced 20 and 40 percent from the levels obtained in Base I, the average land shadow price declines by $2.00 and $6.00 per acre. With the 60 percent reduction in soil erosion, however, the average shadow price increases $21.00 per acre from the Base I alternative. The decline in shadow price for the 20 and 40 percent reduction alternatives reflects the more intensive production on the less erosive land groups and the resulting higher yields. Because of the higher yields, less land is needed. To achieve the 60 percent forced reduction in soil erosion, however, requires the use of more expensive practices on the relatively marginal lands subject to erosion, and the result is the increase in the land shadow prices shown in Table 8.6.

Changes in Land Management Practices

Comparison of the tillage and conservation practices under the various alternatives indicates which land management practices are the most cost effective means of controlling soil erosion losses (Table 8.7). In the Base I solution almost 95 percent of the cropland is under straight-row cropping. This percentage steadily declines to about 45 percent as erosion is reduced to 60 percent of the level obtained in the Base I alternative. The acreage cultivated under contouring increases from 1.5 million acres in the Base I solution to 124.5 million acres when soil losses are reduced by 60 percent. The use of strip cropping and terracing does not increase nearly as much as contouring. These latter two practices are not as cost effective.

Table 8.7. Land use by conservation and tillage practice

Conservation and Tillage Practice	Alternatives			
	Base I	20 Percent Reduction	40 Percent Reduction	60 Percent Reduction
	(million acres)			
Straight Row:				
Fall plowed	13.9	12.4	9.5	6.1
Spring plowed	138.8	107.0	79.5	49.8
Reduced tillage	190.6	184.3	149.1	114.1
Total	343.3	303.7	238.1	170.0
Contour:				
Fall plowed	0.0	0.1	0.3	0.4
Spring plowed	NS[a]	14.2	25.0	27.7
Reduced tillage	1.5	14.5	51.5	96.4
Total	1.5	28.8	76.8	124.5
Strip cropping:				
Fall plow	0.0	0.4	0.7	1.5
Spring plow	3.2	4.6	6.1	4.4
Reduced tillage	3.4	7.1	10.4	5.5
Total	6.6	12.1	17.2	11.4
Terracing:				
Fall plowed	0.5	0.8	1.7	3.6
Spring plowed	10.7	14.1	20.6	28.9
Reduced tillage	2.4	5.2	9.8	24.7
Total	13.6	10.2	32.1	57.2
Grand Total	365.0	364.9	364.4	363.1

[a]NS means not significant.

In the Base I solution, 42 percent of the cropland is spring-plowed and 54 percent is managed under reduced-tillage operations leaving residue on the surface of the fields. The use of fall plowing is quite limited in the Base I solution, only about 4 percent of the total cultivated cropland. Imposing the soil-loss restrictions on the model results in a substitution of reduced-tillage operations for spring plowing. In the 60 percent solution, the percentage of total cropland under spring plowing and reduced tillage is 30 percent and 66 percent, respectively.

Soil erosion losses are also affected by the crops grown and the cropping pattern. As mentioned previously, the use of continuous rowcropping declines with the imposed soil-loss restrictions. The use of rotations with a row crop in only one year out of four, i.e., 25 percent of the rotation sequence is row crop, increases slightly with the 20 and 40 percent erosion control limits (Table 8.8). The increase in the use of rotations that are only 25 percent row crop is significantly larger in the 60 percent erosion control alternative. The crops grown on each land group fluctuate among the various alternatives. The acreage of soybeans on land group I declines while corn and sorghum acreage fluctuates considerably from a 25.7 percent increase under the 20 percent reduction to only a 2 percent increase under the 60 percent reduction. The percentage changes of acreages of crop types does not vary much between alternatives for land group II. Acreages of all three crop types fluctuate widely across the alternatives on land group III. Soybean acreage fluctuates widely on land group IV as the percentage reduction of erosion losses allowed increases. The acreages of all three crop types decline on land group V, the least productive land.

Comparing the soil erosion control alternatives with the Base I alternative to identify differences is straightforward. Determining why the differences occur is more difficult. For example, the use of continuous rowcropping on land group

Table 8.8. Row crop acres by percentage of rotation sequence by alternative and land group

Alternative and Land Group	Percentage of the Rotation sequence that is Row Cropped				
	25	50	75	100	Total[a]
	(thousand acres)				
Base I:					
Land Group I	4,254	15,756	7,803	40,576	68,391
Land Group II	15,422	42,427	19,132	70,955	147,938
Land Group III	4,322	13,117	5,143	27,769	50,532
Land Group IV	2,738	2,046	3,842	5,935	14,563
Land Group V	556	913	1,216	2,300	4,986
Total	27,294	74,262	37,138	147,536	286,232
20 Percent Reduction:					
Land Group I	4,147	20,710	4,616	38,908	68,383
Land Group II	15,263	49,000	15,415	69,078	148,757
Land Group III	4,544	13,987	6,315	25,750	50,597
Land Group IV	2,789	2,602	3,296	5,498	14,187
Land Group V	644	665	1,216	1,612	4,159
Total	27,409	86,966	30,860	140,848	286,084
40 Percent Reduction:					
Land Group I	4,579	21,169	4,063	37,648	67,462
Land Group II	15,335	46,554	18,462	69,377	149,730
Land Group III	4,726	13,886	6,316	26,539	51,468
Land Group IV	2,339	2,931	1,978	6,252	13,502
Land Group V	594	531	1,112	1,498	3,737
Total	27,576	85,074	31,934	141,316	285,901
60 Percent Reduction:					
Land Group I	5,274	21,462	2,583	32,591	61,912
Land Group II	18,508	44,778	20,099	62,953	146,339
Land Group III	5,106	11,777	8,703	31,429	57,017
Land Group IV	1,830	3,286	2,166	7,723	15,007
Land Group V	735	417	761	1,631	3,545
Total	31,455	81,722	34,314	136,329	283,822

[a]Totals incorporate rounding error.

III (a land group with land subject to erosion hazards). (Table 8.1) is higher in the solution with the 60 percent reduction in erosion losses than in the Base I alternative (Table 8.9). This change is possible because of a significant increase in the use of terracing on land group III. To be able to put row crops on the erosive lands under the very stringent soil-erosion restraints requires the protection afforded by terraces. To offset the considerable expense of terracing, the higher-valued row crops must be grown more frequently.

Table 8.9. Crop acreages for small grains, corn and sorghum, and soybeans for the Base and percentage changes for each alternative by land group

Land group and Crop Type	Base I	Alternative: 20 Percent Reduction	40 Percent Reduction	60 Percent Reduction
	(thousand acres)	------(percentage change from Base I)---		
Land Group I:				
Small grains[a]	14,659	-0.8	2.5	5.4
Corn and sorghum	21,687	25.7	22.2	2.0
Soybeans	24,064	-6.6	-11.2	-11.6
Land Group II:				
Small grains	55,834	-5.7	-2.0	-1.4
Corn and sorghum	45,676	-3.8	-0.8	1.6
Soybeans	56,876	2.3	2.9	-5.0
Land Group III:				
Small grains	30,947	-3.1	-7.8	-17.7
Corn and sorghum	10,780	2.4	1.0	27.1
Soybeans	22,646	-49.4	-44.6	-23.9
Land Group IV:				
Small grains	8,408	-2.4	13.7	29.8
Corn and sorghum	4,159	-2.1	-7.6	1.4
Soybeans	1,766	-6.5	-52.0	10.3
Land Group V:				
Small grains	2,826	-14.4	-18.5	-18.9
Corn and sorghum	1,656	-9.8	-28.0	-35.3
Soybeans	949	-60.9	-69.7	-86.3

[a]Includes barley, oats, and wheat.

Impact on Regional Cropping Patterns

A third adjustment possibility to the soil-loss restraint is interregional shifts in cropping patterns, e.g., the shifting of row crops to regions of the country less subject to erosion. Data in Table 8.10 show that most feed grain production occurs in the Lake States, Corn Belt, and the Northern Plains regions under all alternatives. More than one-half of the production of corn, sorghum, barley, and oats is in the Corn Belt region. Imposing the soil-loss restrictions on the model produces only slight changes in the relative share of total feed grain production

Table 8.10. Feedgrain production shares by major zone for each alternative

Major Zone	Base I	20 Percent Reduction	40 Percent Reduction	60 Percent Reduction
	Alternative			
	-----------(percent of total U.S. production)--------			
Northeast	2.5	2.5	2.5	2.7
Southeast	4.0	4.0	4.1	4.1
Lake States	23.8	24.2	24.3	23.8
Corn Belt	51.8	51.3	51.0	51.0
Delta States	0.6	0.6	0.7	1.1
Northern Plains	12.8	12.8	12.9	12.8
Southern Plains	3.7	3.7	3.7	3.8
Pacific	0.8	0.8	0.8	0.8

in each region. Total production of feed grains declines slightly in the Corn Belt region under increasingly stringent erosion controls. Production in the Lake States fluctuates slightly in response to the soil erosion controls.

The slight variations occur because of changes in the comparative advantage among the producing areas for producing feed grain crops. The fluctuations in regional production patterns are small because such a high percentage of the land available in the crop sector is already cropped; 99 percent of the available land in Base I is utilized.

Shadow prices (or supply prices) for feed grains by region are shown in Table 8.11. There is considerable variation among regions for feed grain prices. The price is lowest in the Corn Belt and Lake States. The shadow prices do not vary much from the prices in the Base I alternative in the 20 and 40 percent soil-loss reduction alternatives. However, regional prices increase significantly when the 60 percent soil-loss reduction is imposed on the model.

The regional production shares and crop shadow prices for the other endogenous crops vary in a similar fashion as feed grains. The explanation for the lack of interregional shifts compared to previous studies in this book is the relatively high utilization of the land base and the exogenous livestock sector.

Table 8.11. Feedgrain shadow prices by major zone for each alternative

Major Zone	Base I	20 Percent Reduction	40 Percent Reduction	60 Percent Reduction
	Alternative			
	-------------(dollars[a] per ton)-------------------			
Northeast	99.71	98.90	98.83	111.29
Southeast	103.96	103.90	103.79	123.17
Lake States	75.61	75.60	75.56	91.72
Corn Belt	70.36	70.50	70.36	81.61
Delta States	114.06	113.19	106.82	112.30
Northern Plains	104.17	104.95	104.62	122.75
Southern Plains	132.68	133.98	133.30	166.40
Pacific	104.47	146.30	145.27	172.09
United States	81.19	81.47	81.39	96.03

[a]These prices are in 1975 dollars.

Consequences for Landowners

Changes in values imputed by the model to land resources under the various alternatives provide valuable information about the potential effects of the soil-erosion controls on the owners of U.S. agricultural lands. Comparing the land values in the alternatives to those in Base I shows the differential impacts of the soil-erosion restraints. When the land values are compared at the national level by land group, there is a significant decline in the value for land group V relative to the other land groups. This implies that the owners of these less productive, marginal lands will be adversely affected by soil erosion control policies.

On a regional basis, imposing the 20 percent reduction soil-loss restraints has little impact. Some slight differential effects occur under the 40 percent restraint, as land values in the Pacific region do not fall while the values in all other regions do. Under the 60 percent reduction alternative the changes are more marked. Land values in all regions rise but the increases in the Southeast, the Corn Belt, and Delta States are relatively small. The land values in the Northern and Southern Plains and the Pacific regions rise markedly more than the average for the nation. The implication of these changes in land values is that control of soil erosion will result in interregional shifts of wealth of landowners, as represented by relative changes in land values.

ble 8.12. Cost of land by land group and alternative for the United States and the eight major zones

ternative and nd Group	United States	Northeast	Southeast	Lake States	Corn Belt	Delta States	Northern Plains	Southern Plains	Pacific
	(dollars per acre)								
se:									
Land Group I	157	183	163	146	182	149	137	127	167
Land Group II	137	178	158	130	182	152	102	101	142
Land Group III	111	175	154	126	168	148	90	95	103
Land Group IV	107	174	148	125	172	152	92	92	104
Land Group V	111	170	132	117	154	154	80	79	94
Average	135	178	159	132	180	152	102	105	139
Percent Reduction:									
Land Group I	176	208	179	164	182	168	178	153	229
Land Group II	156	203	160	150	191	168	173	124	196
Land Group III	133	202	178	144	181	167	116	120	114
Land Group IV	124	199	171	128	164	155	115	110	135
Land Group V	99	187	88	76	93	122	102	87	108
Average	156	204	168	152	186	165	128	128	184

RESULTS FOR AN EXPANDED AGRICULTURAL LAND BASE AND ALTERNATIVE PROJECTIONS OF CROP YIELDS FOR 2030

As explained previously, the commodity demands imposed in the Base I alternative require almost all of the available cropland to be put into production. The alternatives considered in this analysis are formulated to examine the extent to which it is economical to convert pasture and forestlands to cropland while varying the assumed rate at which crop yields will increase over the next 50 years.

When the 127.7 million acres of medium and high potential croplands are available to utilize in the Base II alternative, the model determines that the commodity demands can be met at a lower cost by converting some of the potential lands to cropland than by cropping the already available cropland as intensively as in Base I.

Table 8.13. Aggregate values utilization of the agricultural land base

Alternative	Available Cropland	Dry Cropland	Irrigated Cropland	Percent of Cropland Utilized[a]
	(million acres)			
Base I[b]	388.7	337.3	27.7	99
Base II[c]	493.4	342.7	29.2	80
Low Crop Yields	493.4	382.4	33.3	89
High Crop Yields	493.4	280.6	23.2	66

[a]Includes 23.5 million acres of exogenous cropping in the calculation.

[b]Includes the 23 million acres of wetland potentially available for conversion to cropland (Table 8.3).

[c]Includes the 127.7 million acres of medium- and high-potential cropland from the 1977 NRI. This includes the 23 million acres in footnote b above.

This high utilization of available cropland in Base I indicates that the capacity of U.S. agriculture will be pressed in the year 2030 if only the 23 million acres of wetland shown in Table 8.3 is available to be converted. Allowing the conversion of additional lands included in the NRI medium and high potential categories markedly changes this result. The percent utilization of the cropland base declines to 80 percent in the Base II alternative.

Varying the rate of improvement in crop yields provides additional information about the adequacy of the resource base for agriculture in the year 2030. Even with the additional 127.7 million acres available for conversion, if crop yield improvements are less than projected, the utilization of the land base approaches 90 percent. However, if crop yields in 2030 should be as high as in the high crop yield alternative, only about two-thirds of the land base will need to be cropped.

Crop Yield and Price Changes

The yields for selected crops presented in Table 8.14 for the three alternatives are averages over the whole U.S. agricultural sector. When yields higher than those projected in Base II are incorporated in the crop sector, fewer acres are required to meet the commodity demands in the model. Thus some of the less productive land is not needed. The result is that average yields are even higher because fewer marginal acres are cropped. The opposite situation occurs when lower yields are projected.

Table 8.14. Average yields, total acreages, and supply prices for selected commodities

	Base II			Low Crop Yields			High Crop Yields		
	Yields[a]	Acreage[b]	Prices[c]	Yields	Acreage	Prices	Yields	Acreage	Prices
Corn	137	66,177	1.00	123	76,296	1.30	167	56,585	.70
Wheat	42	84,514	1.70	38	94,279	2.10	54	67,150	1.30
Soybeans	46	92,081	4.20[d]	40	104,596	5.60[d]	60	70,209	2.90[d]

[a]Yields for corn, wheat, and soybeans are in bushels per acre.

[b]Thousands of acres.

[c]Prices for corn and wheat are in 1975 dollars per bushel (for soybeans see footnote d).

[d]These are dollars per hundred weight for oilmeal in 1975 dollars.

Aggregate Resource Use

Considering only the three alternatives which include the 127.7 million additional acres as potentially available for conversion and using the Base II alternative as the point for measuring the impact of varying crop-yield possibilities on agriculture, the results for selected variables are shown in Tables 8.13 and 8.14. The potential impact on consumers is shown in Table 8.15. The cost of production fluctuates markedly. The impact of changing the productivity of the whole agricultural sector is indicated by the changes in total land cropped, as well as in other agricultural inputs. The use of pesticides and herbicides fluctuates more widely across alternatives than nitrogen fertilizer use.

Table 8.15. Indices for variables aggregated to the national level for the alternative yield solution

Alternatives	Food and Fiber Cost of Production	Soil Erosion	Nitrogen	Pesticides and Herbicides
	(Base II = 100)			
Base II	100	100	100	100
Low Crop Yields	123	116	110	115
High Crop Yields	74	83	83	73

THE 1985 RESOURCES CONSERVATION ACT EVALUATION

With the 1980 RCA analysis completed, attention is now on the 1985 RCA evaluation. The analysis to be carried out at CARD for the 1985 RCA will employ an interregional linear programming model. This model will be an updated and modified version of the model employed in the 1980 RCA effort. The modifications include making the livestock sector endogenous, adding a range-pasture sector, expanding the crop sector with an additional tillage option and two additional crops, and redefining the land groups in the land-base restraints.

In addition, the 1985 RCA analysis will include solutions with endogenously determined demand quantities and prices. A national econometric model incorporating commodity and resource demand equations and certain structural components will be combined in a recursive fashion with a national and regional programming model. In some cases, e.g., to examine possibilities of targeting conservation funds, a national supply-demand econometric model will be combined with a regional land-water model that incorporates supply sectors for commodities and soil loss.

REFERENCES

Allen, George C., and Margaret Devers. 1975. Livestock-Feed Relationships, National and State. Supplement for 1974 to Statistical Bulletin 530. USDA, Economic Research Service, Washington, D.C.

Colette, Arden W. 1974. The Conceptualization and Quantification of the Water Supply Sector for a National Agricultural Analysis Model Involving Water Resources. Center for Agricultural and Rural Development, Iowa State University, Ames.

Dyke, Paul, and Linda Hagen. 1980. Soil Loss and Its Effects on Yields: Functional Relationships Estimated. USDA, Economics, Statistics, and Cooperatives Service, Washington, D.C. Unpublished data.

English, Burton C., Klaus Alt, and Earl O. Heady, 1982. A Documentation of the Resources Conservation Act's Assessment Model of Regional Agricultural Production, Land and Water Use, and Soil Loss. Miscellaneous CARD Report 107T. Center for Agricultural and Rural Development, Iowa State University, Ames.

Eyvindson, Roger H. 1965. A Model of Interregional Competition in Agriculture Incorporating Consuming Regions, Producing Areas, Farm Site Groups, and Land Classes, volumes 1-5. Unpub. Ph.D. diss., Iowa State Univ., Ames.

Ibach, D. B., and J. B. Adams. 1976. Fertilizer Use in the United States by Crops and Areas, 1964 Estimates. Statistical Bulletin 408. USDA, Economic Research Service and Statistical Reporting Service, Washington, D.C.

Meister, Anton D., and Kenneth J. Nicol. 1975. A Documentation of the National Assessment Model of Regional Agricultural Production, Land and Water Use, and Environmental Interaction. CARD Miscellaneous Report. Center for Agricultural and Rural Development, Iowa State University, Ames.

Nicol, Kenneth J., and Earl O. Heady. 1975. A Model for Regional Agricultural Analysis of Land and Water Use, Agricultural Structure, and the Environment: A Documentation. CARD Miscellaneous Report. Center for Agricultural and Rural Development, Iowa State University, Ames.

Short, Cameron, and Dan Dvoskin. 1977. Utilizing Animal Waste as a Source of Nitrogen. CARD Miscellaneous Report. Center for Agricultural and Rural Development, Iowa State University, Ames.

Stoecker, Arthur. 1974. A Quadratic Programming Model of U.S. Agriculture in 1980. Theory and Application. Unpub. Ph.D. diss., Iowa State Univ., Ames.

U.S. Department of Agriculture, Agricultural Resource Assessment System Technical Committee. 1975. Unpublished working papers prepared for the National Water Assessment. Statistical Reporting Service, Washington, D.C. Mimeographed.

U.S. Department of Agriculture, Conservation Needs Inventory Committee. 1971. National Inventory of Soil and Water Conservation Needs, 1967. Statistical Bulletin 461. Washington, D.C.

U.S. Department of Agriculture, Economic Research Service. 1976. 1973 and 1974 Crop Budgets. Firm Enterprise Data System, Oklahoma State University, Stillwater.

___.1977. 1975 Crop Budgets. Firm Enterprise Data System, Oklahoma State University, Stillwater.

U.S. Department of Agriculture, Soil Conservation Service. 1980. 1977 National Resource Inventories. Washington, D.C.

U.S. Department of Agriculture, Soil Conservation Service, Special Projects Division. 1976. Crop Consumptive Irrigation Requirements and Irrigation Efficiency Coefficients for the United States. West Technical Service, USDA, Portland, Oregon.

U.S. Department of Commerce, Bureau of Census. 1977. *U.S. Census of Agriculture, 1974.* Vol. I: *Statistics for the States and Counties*. U.S. Government Printing Office, Washington, D.C.

U.S. Water Resources Council. 1970. Water Resources: Regions and Subareas for the National Assessment of Water and Related Land Resources. U.S. Government Printing Office, Washington, D.C.

___.1975. 1972 OBERS Projections, Regional Economic Activity in the U.S. Agricultural Supplement. U.S. Government Printing Office, Washington, D.C.

Wischmeier, W. H., and D. D. Smith. 1965. Predicting Rainfall-Erosion Losses from Cropland East of the Rocky Mountains. USDA Agriculture Handbook 282. Washington, D.C.

SECTION V. INCORPORATING DEMAND RESPONSE IN ENVIRONMENTAL MODELS OF U.S. AGRICULTURE

A LIMITATION of conventional linear programming models is the assumption of fixed commodity demands implied in the linear objective function. Approaches used to relax this assumption include quadratic programming models and various linear approximation techniques (separable programming) that allow incorporation of demand relationships in the programming model. Also it is possible to combine econometric demand-oriented models and programming models. The advantage of incorporating demand relationships in the programming model is that equilibrium commodity prices and quantities are determined simultaneously in the model.

The studies in this section all employ models which directly incorporate demand relationships in the analysis. The first study utilizes an interregional quadratic programming model of the U.S. agricultural sector to evaluate impacts of two potential environmental-protection policy alternatives. One policy alternative limits the rate of nitrogen fertilization in U.S. agriculture. A second policy alternative removes four organochlorine insecticides from the market. The two alternatives imply a free market except for restrictions on the use of nitrogen fertilizer and the specified insecticides. The effect of these policy alternatives are compared with a third alternative that does not have restrictions on fertilizer and insecticide use. The price effects caused by limiting the fertilization rate are more substantial than the price effects resulting from the insecticide removal. The nitrogen restriction results in price increases for most commodities because of lower yields. The price increases were moderated by lower consumption of agricultural products due to the demand relationships in the model.

The second study utilizes an interregional separable programming model to analyze the potential of alternative policies to simultaneously increase farm income and decrease soil erosion. Four supply control and soil conservation policies are analyzed for their impact on regional farm prices and incomes, farm output, farm input use, and gross soil erosion. The four policy alternatives are a baseline alternative that does not include a land retirement or soil-loss policy; a soil conservation alternative in which a maximum allowable per acre soil-loss restraint is imposed on U.S. agriculture; a land retirement, supply control policy alternative designed without special conservation aspects; and a land retirement policy alternative designed as a simultaneous supply control and soil conservation program. The analysis indicates that a conservation-oriented land retirement policy can be designed to increase net farm income while simultaneously achieving significant reductions in gross soil erosion.

The third study in this section describes a modeling system that is being constructed in cooperation with the International Institute of Applied Systems Analysis (I.I.A.S.A.), which incorporates a programming model within the framework of an econometric simulation model. The specific objective of the study is to evaluate both the regional and national impacts of selected legislative policies of the state of Iowa aimed at controlling soil erosion. A regional programming model of Iowa agriculture is linked recursively with a U.S. national econometric simulation model. The focus is on the state of Iowa with necessary attention given to Iowa's position within the agricultural economy of the United States. Results from the modeling system are presented, as well as a discussion of the plans for the improvement and future uses of this model.

CHAPTER 9. AN INTERREGIONAL COMPETITION QUADRATIC PROGRAMMING ANALYSIS OF TWO ENVIRONMENTAL ALTERNATIVES FOR U.S. AGRICULTURE

by Kent D. Olson, Earl O. Heady, Carl C. Chen, and Anton D. Meister

THIS STUDY of two environmental policy alternatives for U.S. agriculture employs an interregional quadratic programming model to determine the impacts of the policies. The first policy alternative investigates limits on the rate of nitrogen fertilization in U.S. agriculture. The second policy alternative in the study removes four organochlorine insecticides from the market. The interregional quadratic programming model allows investigation of the interregional competition aspects of these environmental policies. In addition, it is possible to evaluate the consumer response to changes in agricultural production costs due to the environmental policies because the demand functions are directly incorporated in the model, allowing both prices and quantities of the agricultural commodities to be determined simultaneously.

The remainder of this introductory section provides a brief explanation of the rationale for the environmental policies under investigation. The following two sections provide a brief mathematical statement concerning the theory of quadratic programming and a detailed explanation of the development of the model employed in this study.

Nitrogen fertilizer is considered by some to be a problem pollutant when it enters water supplies, raising the concentration of nitrate ions in the water. Nitrogen fertilizer may enter water supplies in two forms. As a nitrate ion, nitrogen can be leached from the application site by percolating water, and as an ammonium ion attached to soil particles, nitrogen can be removed from the field and enter the water via eroded soil. A high nitrate ion concentration is a concern for humans; the disease infant methemoglobinemia is caused by water nitrates. Also, hogs and cattle exhibit poor growth characteristics with nitrates in their water. This study does not intend to prove or disprove these concerns, but it is intended to show economic impacts of government policy that might be used to alleviate these potential problems.

Insecticides allow crop production in insect-infested areas. This study examines the production effects of removal of four organochlorine insecticides–Aldrin, Dieldrin, Chlordane, and Heptachlor–from agricultural use. At the time of this study, the use of Aldrin and Dieldrin had already been suspended by the Food and

Drug Administration (FDA) and Chlordane and Heptachlor were being considered for suspension. Persistence, or length of active life in the soil, is the major environmental problem with these insecticides. The substitutes employed in this analysis, three organophosphate insecticides and one carbamate insecticide, have a shorter persistence in the soil. These substitutes, however, are not as effective against some important pests. The analysis is concerned with the impacts of the higher-cost substitutes and their ineffectiveness against some insect pests.

THEORY OF QUADRATIC PROGRAMMING

Defining a primal linear programming problem as minimization of production costs subject to resource constraints and minimum production levels, the optima will be denoted as $f(\bar{x})$. The dual of this problem is the maximization of gross producer profits subject to the restriction that net profit of each activity be zero or negative. (Gross producer profits are used here as the sales of goods minus purchases of resources.) Let the optimal solution of the dual be $g(\bar{w}, \bar{u})$ where w is the vector of imputed prices of the goods and u is the vector of imputed values of resources. From basic linear programming theory we know that if firms face the prices $\bar{w}$ and $\bar{u}$, the quantities $\bar{x}$ will be produced, and given quantities $\bar{x}$, prices $\bar{w}$ and $\bar{u}$ will result.

Combining the primal and dual problems, the problem is now to maximize net producer profit subject to resource constraints, minimum production levels (i.e., supply $\geq$ demand), nonpositive pure profit, and the usual nonnegativity constraints, as were assumed above implicitly. These constraint equations are skew symmetric, thus making the feasibility space that Tucker (1956) and Goldman and Tucker (1956) refer to as self-dual. Denoting the optima as $(\bar{x}, \bar{u}, \bar{w})$, it can be shown that these optimal values are the same as those from the primal and dual problems.

In the primal problem, the minimum production levels can be considered as levels of demand determined outside the model. In the dual problem, the production levels can also be described as the levels of demand. If demand is now described as a linear function of prices (9.1), a quadratic objective function is evident. Letting small letters denote vectors and capital letters denote matrices:

$$d = d_o + Dw \tag{9.1}$$

where d = a vector of demands at imputed prices; d_o = a vector of given demands (intercepts); D = a negative semidefinite matrix of linear own and cross demand slopes (D is not required to be symmetric); and w = a vector of imputed prices. Plugging (9.1) into the primal linear problem and using Hanson's (1961) duality theorem, we can obtain a quadratic problem that is self-dual, i.e.:

$$\text{Maximize } \emptyset\ (x,u,w) = d'_o w + w'Dw - b'u - c'x \tag{9.2}$$

$$\text{subject to } DW \quad -Ax \leq -d_o \tag{9.3}$$

$$Bx \leq b \tag{9.4}$$

$$A'w - B'u \leq c \tag{9.5}$$

$$w,u,x \geq 0 \tag{9.6}$$

where d_o, D, and w are as previously defined. A and B are matrices of technical coefficients that describe the transformation of each of the primary resources through the production activities into a set of final quantities demanded. The vector c contains the exogenous costs associated with each of the production activities. The vector b is the available primary resources while u is the value imputed to these resources. Vector x is, of course, the calculated levels of production of the various activities. The constraint matrix is skew symmetric for the matrix D so we have a quadratic self-dual system.

This objective function (9.2) maximizes net producer profit. Because of the restraints put on the model, this net producer profit will be optimal at zero. This is apparent, since we know that if marginal cost (MC) is greater than marginal revenue (MR), no activity will come into the solution, and if MR is greater than MC, the activity can be increased until MR = MC. Constraint (9.5) says that the value of an activity cannot exceed the exogenous cost of that model plus the imputed value of the primary resources used in that activity.

Constraint (9.3) allows supply to be equal to or greater than demand, but not less than demand. Constraint (9.4) puts a limit on the amount of primary resources available. Constraint (9.6) is the normal nonnegativity requirement. Taken together, constraints (9.3) to (9.6) describe the equilibrium conditions in a competitive market.

A QUADRATIC PROGRAMMING MODEL FOR U.S. AGRICULTURE IN 1980

In this section, a national quadratic model for U.S. agriculture in the year 1980 is developed. The model in this study draws on the work of Plessner (1965), Hall (1969), Stoecker (1974), and Chen (1975). The regional delineation of the model will be defined first, and then a detailed explanation of the specification of the model is provided. For further details concerning the model, see Olson et al. (1977).

Regional Delineation of the Model

The forty-eight continental states and the District of Columbia are divided into 10 spatially separated consuming regions (Figure 9.1). These 10 consuming regions are further subdivided into 103 producing areas (Figure 9.2). The seventeen western states are divided into 10 irrigated crop-producing areas (Figure 9.3), which differ from those of previous models and studies in this book.

Crop production is defined at the producing-area level and, where appropriate, at the irrigated-area level. Livestock production is defined at the consuming-region

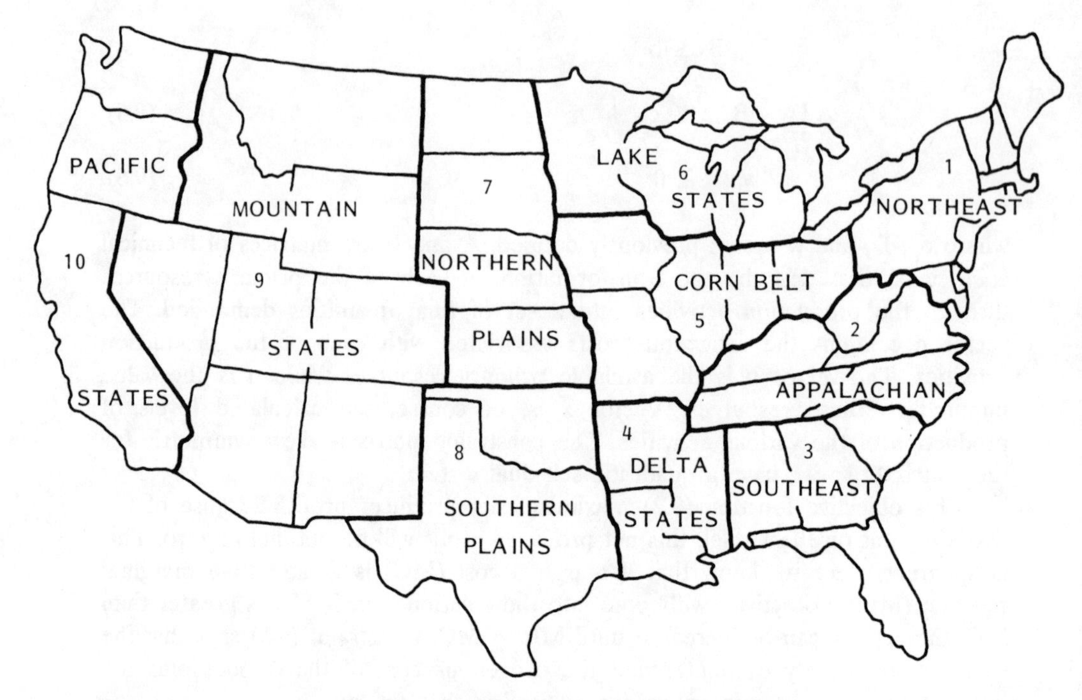

Figure 9.1. Location of the 10 consuming regions and livestock-producing regions.

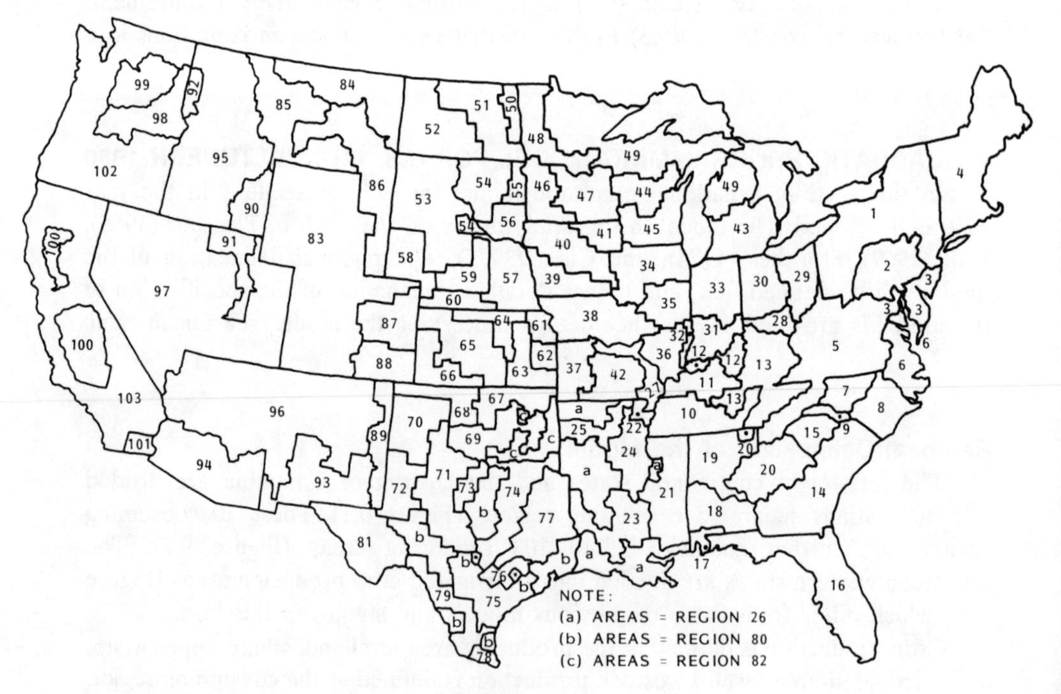

Figure 9.2. The 103 crop-producing areas.

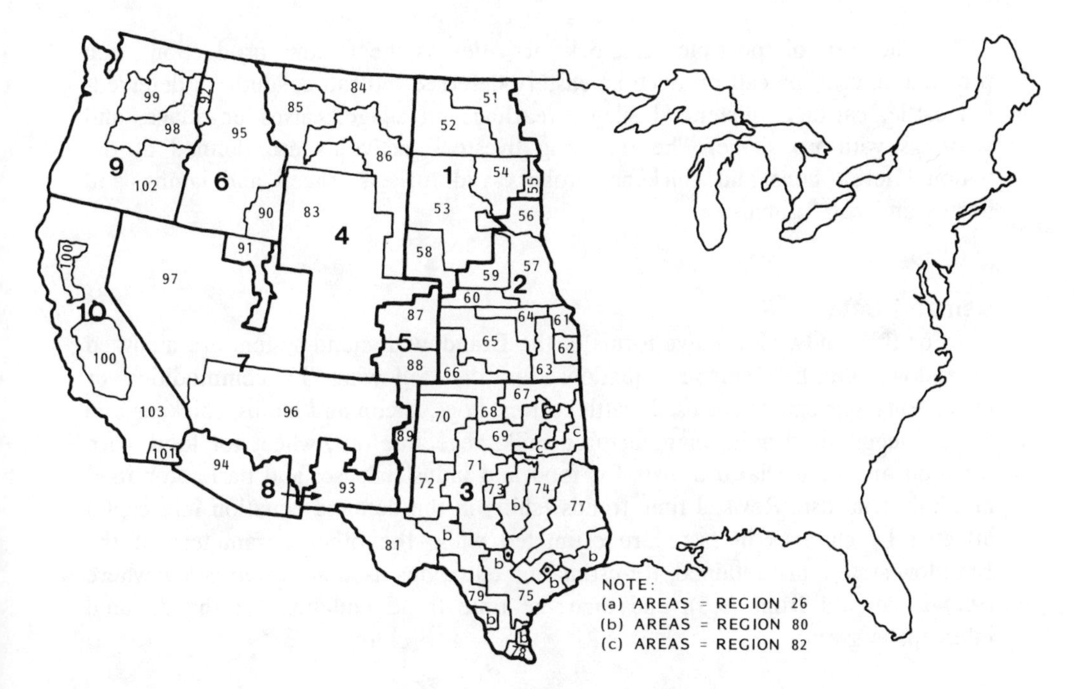

Figure 9.3. The 10 irrigated crop-producing areas.

level. Producers of a commodity within an area or a region are assumed to be homogeneous with respect to technology.

Transportation is defined among the 10 consuming regions for specific final and intermediate commodities. It is assumed that corn, oats, and barley for food are perfect substitutes for corn, oats, and barley for feed, respectively, and vice versa. Wheat can also be used as a feed source. Demand can be satisfied by production within a region and (or) through commodities shipped from outside the region. Feed exogenous to the model can be purchased by the appropriate activity in the model. Inputs exogenous to the model are considered to be unlimited in quantity at a given, set price.

Definition of Activities

A crop activity is defined for a producing area if 1000 acres or more of that crop were reported in that area in 1969. The set of possible crop activities includes wheat, corn, oats, barley, feed grain (corn, oats, barley, grain sorghum), feed grain-soybean rotation, feed grain-hay rotation, feed grain-silage rotation, and hay-silage rotation. Irrigated crop activities are defined similarly. If cotton has been grown in a consuming region, a cotton production activity is defined for that region.

A livestock activity is defined for a consuming region if 1000 or more units of that activity were reported in that region on an annual basis between 1959 and

1978. The set of possible livestock activities is beef cow production, hog production, yearling calf production, eastern deferred-fed cattle, southern deferred-fed cattle, cattle on extended silage, yearlings on silage, calves on silage, and yearlings with no silage. The following livestock activities are defined at the national level: hens and chickens, broilers and turkeys, sheep and lambs, and horses and zoo animals.

Demand Data

For this study, alternative forms of the Brandow demand system are analyzed (Brandow, 1961). Demand equations for the following 13 commodities or commodity aggregates are used: cattle, calves, hogs, sheep and lambs, chickens and turkeys, eggs, fluid milk, manufactured milk, vegetable oils, wheat for food, corn for food and industrial use, oats for food and industrial use, and barley for food and industrial use. Revised time trends (shifts in the demand-equation intercepts) affected by changes in taste are estimated while the other parameters of the Brandow system are retained. Re-estimation equations used are given below where (9.7) is derived from (9.8), and then the time trend equation for the demand intercept is given:

$$q_{it} = d_{it} + D_iP_t \tag{9.7}$$

$$d_{it} = q_{it} - D_iP_i = a_0 + a_1T + e_t \tag{9.8}$$

where q_{it} = total quantity of the ith commodity demanded in year t; d_{it} = demand intercept of commodity i in year t; D_i = ith row of the demand matrix; and D_iP_t = set of prices, consumer income, and the index of nonfood prices in year t. Ordinary least squares is the estimation procedure used. When the Durbin-Watson statistic shows autocorrelation, a one-step autocorrelated error model is used.

For the demand matrix itself, Stoecker (1974) describes the method of selecting the variation of Brandow's system. Briefly, three algebraic forms of the demand equations were viewed: (1) constant elasticities, (2) Brandow's slopes, and (3) Hall's (1969) slopes. Alternatives in two other areas were also looked into: nominal vs. deflated farm-level prices and constant total farm-level demand slopes vs. constant per capita demand slopes.

Comparisons among results from the variations were based on the Theil's U statistic, the standard error of the equation, the average relative error, and the absence of first order autocorrelation. The per capita form of the demand equation was employed because of its greater consistency with the idea of the representative consumer and because its performance is slightly better than the total market demand forms. Nominal prices generally performed better than deflated prices.

The constant elasticity forms of the Brandow equation were the better of the three alternatives. However, linear demand equations are needed in the constraint set, so these were based on converting Brandow's elasticities to slopes using 1963-

65 average prices and quantities. The national demand matrix and intercepts, corrected for exports, are given in Table 9.1.

This national demand matrix is partitioned on the basis of population into regional demand matrices for 10 commodities. Demand for other desired commodities is defined as follows: cotton lint demand is fixed at a national level; demand equations for chickens and turkeys, eggs, and sheep and lambs are specified on the national level.

The regional demand matrix is partitioned into the submatrices below:

$$B'_k = \left[\begin{array}{c|c} D_{rk} & C_k \\ \hline R_k & D_{nk} \end{array}\right] \tag{9.9}$$

where B'_k = 13 × 13 matrix of demand slopes for consuming region k = 1,2, ..., 10; D_{rk} = 10 × 10 matrix measuring the effect on regional demand in terms of regional prices; C_k = 10 × 3 matrix relating the effect of national prices to quantities demanded in region k; R_k = 3 × 10 matrix relating the effect of prices in region k to national demands; and D_{nk} = 3 × 3 subregion demand matrix. Summation of D_{nk} over k equals D_n. C_k, R_k, and D_{nk} are necessary due to the specification of demand of 3 commodities on the national level.

The regional demand matrices, B'_k, are derived from the national demand matrix by the following relationship:

$$B'_k = w_k{*}D \tag{9.10}$$

where w_k = the proportion of total population in the kth region, and D = a national demand matrix (Table 9.1).

The regional demand intercepts are derived in a manner similar to the regional slopes, but the intercepts are also adjusted for expected regional differences in personal disposable income:

$$d_k = w_k\,[d + I_d\,(I_k - I_{us})] \tag{9.11}$$

where new terms are defined as d_k = regional demand intercept; d = national demand intercept; I_d = regional factor relating changes in personal disposable income to the quantity demanded at the national level; I_k = expected personal disposable income per capita for the kth consuming region; and I_{us} = expected personal disposable income per capita for forty-eight states and the District of Columbia.

Domestic demand for cotton lint is set at 17 pounds per capita, or 8.1 million bales. Net commercial export of cotton is set at 4.2 billion bales. Total demand for cotton lint is thus 12.3 million bales.

Demand for a desired commodity is allowed to be satisfied by production in any region, using the available transportation activities. Production within a consuming region satisfies that region's demand with no transportation costs.

Table 9.1. National farm-level demand for food use in 1980[a]

Commodity[b]	Cattle	Calves	Hogs	Fluid Milk	Mfd. Milk	Oil	Wheat	Corn	Oats	Barley	Sheep & Lambs	Chickens & Turkeys	Eggs	Intercept
CA	-1,300.260	67.598	124.640	0.048	3.983	0.125	2.657	1.066	0.167	0.024	60.329	2.965	149.687	72,822.8
CF	26.636	-97.912	11.853	0.002	0.205	0.007	0.137	0.055	0.009	0.001	5.569	0.153	14.096	1,438.4
HG	109.692	26.200	-559.373	0.029	2.388	0.075	1.593	0.639	0.100	0.041	31.532	1.778	79.231	25,268.4
FM	2.092	0.322	1.069	20.902	1.678	0.017	0.085	0.034	0.005	0.001	0.194	0.269	0.909	5,209.1
MM	4.787	0.739	2.448	1.611	-84.624	4.503	0.458	0.184	0.028	0.004	0.446	0.560	2.081	8,671.2
OL	0.437	0.068	0.224	0.030	3.544	-1.845	0.034	0.014	0.002	0.001	0.041	0.042	0.190	2,286.4
WH	6.619	1.023	3.387	0.450	1.233	0.045	-9.892	0.730	0.114	0.016	0.618	0.865	2.878	14,703.0
CN	4.336	0.670	2.219	0.295	0.809	0.030	1.194	-13.940	0.075	0.011	0.405	0.566	1.886	4,757.5
OT	0.597	0.092	0.305	0.041	0.111	0.004	0.164	0.066	-2.563	0.002	0.056	0.078	0.260	590.5
BY	0.067	0.010	0.035	0.005	0.013	0.001	0.019	0.007	0.001	-9.297	0.006	0.009	0.029	1,401.1
SL	43.197	10.319	26.258	0.002	0.203	0.006	0.135	0.054	0.009	0.001	-181.712	0.151	21.736	3,123.1
CT	3.732	0.577	1.910	0.254	0.578	0.016	0.294	0.118	0.019	0.003	0.349	-40.394	1.623	6,733.5
EG	64.863	17.881	45.512	0.013	1.056	0.033	0.703	0.282	0.044	0.006	15.156	0.785	-466.177	17,650.7

[a]Commodities are: cattle, calves, hogs, fluid milk, manufactured milk, oil, sheep and lambs and poultry meat in cwt; wheat, corn, oats, and barley in bushels; eggs in hundred dozen. All prices are expressed in 1963-65 dollars. Quantity changes are in 10,000 units. The slope coefficients showing the effect of one-unit changes in the farm price of each commodity and the domestic intercept term for the commodities are listed row by row.

[b]Commodity code: CA, cattle; CF, calves, HG, hogs; FM, fluid milk; MM, manufactured milk; OL, oil; WH, wheat; CN, corn; OT, oats; BY, barley; SL, sheep and lambs; CT, chicken and turkey; EG, eggs.

Exports

In this model exports are defined as net exports (i.e., total commercial exports less imports). Estimates of net exports were made for all the final commodities plus feed grains and oilmeal. These estimates of 1980 foreign demand are either a fixed amount or based on a linear equation involving an intercept and an inverse relation to the commodity's own price. In estimating the export levels, time trend was used to give "normal" exports. These normal exports are lower than the export levels of the past few years. (They do not conform to future exports presented in the previous table.) For wheat, the estimated export demand for 1980 is 1000 million bushels compared to a 1968-70 average of 628 million bushels and a 1972-73 average of 1165 million bushels. Corn exports for 1980 are set at 950 million bushels; this compares to 553 and 1250 million bushels for 1968-70 and 1972-73 average exports, respectively. Allocation of net exports among ports, and thus consuming regions, is made from historical patterns of shipment. These regional equations or intercepts are added to the appropriate rows in the demand matrix.

Exogenous Feed Demands and Supplies

The model includes only the major types of livestock production as endogenous activities. Feed and pasture needs for exogenous livestock production have been treated as fixed negative supplies in the appropriate rows of the demand intercept vector. The crop activities in the model do not produce all the feed supplied to the livestock industry. Fishmeal, linseed meal, rice millfeeds, corn gluten meal, wheat bran and middlings, and brewers' byproducts, for example, are available from various nonagricultural sources and thus exogenous to the model.

Land Base and Rotation Weights

Land resources in the continental United States are categorized as cropland, cropland plus hay land, irrigated cropland, irrigated cropland plus hay land, wild hay land, pasture, and cotton land. Irrigated cropland and irrigated cropland plus hay land are defined similarly to nonirrigated land except the acreages are adjusted to include land brought under irrigation through 1969 and estimates of new irrigated land from Bureau of Reclamation projects scheduled for completion by 1980 (Heady et al., 1972).

Wild hay land is the 1953 harvested acreage of wild hay. The year 1953 is used because it is the last year in which acreages were not significantly affected by government programs. Pasture is measured in animal unit months (AUM) available for livestock production in each of the 10 consuming regions. Pasture includes woodland pasture, permanent pasture, improved permanent pasture, cropland pasture, unimproved permanent pasture, and aftermath pasture. In addition, all land resources, except wild hayland, are assumed to produce pasture if not used for crop production. Thus, each consuming region has a total AUM figure that is decreased as crop production takes place in that region.

If each crop is defined singly in a cropping activity, the programming model has a tendency to produce only one crop, the most profitable crop, in each area. This result does not reflect the actual practices of farmers, as crop rotations are typically employed to reduce risks, protect the soil, distribute labor requirements over the cropping season, etc. To overcome this tendency, activities are defined to give combinations of crops in rotations. Feed grains is an example where these activities are developed. The relative proportions (rotation weights) of each of these individual crops in the feed grain activity was based on total acreages of each crop in 1964 and 1969 in each producing area. These historical weights are defined for the following rotational activities: feed grains, feed grain-soybeans, feed grain-silage, feed grain-hay, and hay-silage.

Crop Yield Projections

Crop yields for 1980 not available at the time of the study were projected on the basis of historical trends adjusted for change in the proportion of acreage under irrigation and for changes in fertilizer practices. Fertilization practices analyzed are the proportion of the crop acreage receiving fertilizer and the quantity of fertilizer applied per acre fertilized.

For each state and crop, response to fertilization was assumed to be given by a single variable Spillman function:

$$Y_t = Y_t^0 + A(1 - R^{x_t})P_t \qquad (9.12)$$

where Y_t = estimated average per acre yield in year t; Y_t^0 = estimated average per acre yield on unfertilized land in year t plus other effects described shortly; A = potential response obtainable from fertilization and is assumed constant; R = 0.8 for all crops; x_t = optimal quantity of fertilizer applied to an acre of land in year t; P_t = proportion of acreage receiving fertilizer in year t; and t = years after 1949.

In the normal Spillman formulation, R is the ratio of successive marginal products. Ibach and Adams (1968) suggest holding R constant at 0.8 for all crops and redefining the unit of fertilizer. This redefinition consists of dividing the total poundage of elemental nitrogen, phosphorus, and potassium by a factor which Ibach and Adams obtained by regression.

Since A and R are held constant, the term $A(1 - R^{x_t})$ represents the response from fertilization only. The response to factors other than fertilization must then be in the term Y_t^0, where Y_t^0 can be defined in terms of (9.12) or as a simpler linear time regression:

$$Y_t^0 = Y_t - A(1 - R^{x_t})P_t = a_0 + a_1T + e_t \qquad (9.13)$$

where all terms are as described previously and T is a time variable with T for 1969 = 0. Thus Y_t^0 is the average per acre yield on unfertilized land plus the yield increases due to technical changes over time other than fertilization rate changes.

The profit maximizing application rate of fertilizer is given in equation (9.14).

$$x_t^* = \frac{\ln [(P_{x,t-1})/(P_{c,t-1})] - \ln A - [\ln (-\ln R)]}{\ln R} \tag{9.14}$$

where x_t^* = optimal number of units of fertilizer to be applied; ln = natural logarithm operator; $P_{c,t-1}$ = price of a unit of crop c, lagged one year; $P_{x,t-1}$ = weighted price of one unit of fertilizer (as redefined), lagged one year; R = 0.8; and A = potential response from equation (9.11).

Because changes in the factor/product price ratio do not account for all of the increased applications of fertilizer, these increases are viewed as an adjustment to the optimal level. Projected fertilizer application rate ratios are based on equation (9.15):

$$\frac{x_t^a}{x_t^*} = q_0 + q_1 T + w_t, \; q_1 \geq 0 \tag{9.15}$$

where x_t^a = estimated actual rate of application in year t.

The projected proportion of acreage receiving fertilizer is based on a linear time trend analysis:

$$P_t = f_0 + f_1 T + u_t \tag{9.16}$$

Crop yield projections at the state level for 1980 (9.17) under specified prices are made by evaluating equations (9.13) through (9.16) for t = 16 (i.e., 1980-1964 = 16):

$$Y_{80}^0 = a_0 + 16a_1 \tag{9.13a}$$

$$X_{80} = X_{80}^* \text{minimum } (1.0, q_0 + 16q_1) \tag{9.14a and 9.15a}$$

$$P_{80} = \text{minimum } (1.0, f_0 + 16f_1) \tag{9.16a}$$

$$Y_{80} = Y_{80}^0 + A(1 - R^{x80}) P_{80} \tag{9.17}$$

Stoecker (1974) described the changes in the above procedure that are made to evaluate the irrigated and nonirrigated yields in the seventeen western states.

A Spillman production function was specified for each crop activity defined in the 103 producing areas and the 10 irrigated producing areas. This was done by aggregating equation (9.14) over the Ibach and Adams' (1968) subregions that intersect with the producing area and by aggregating equations (9.13), (9.15), and (9.16) over the states that intersect with the producing area.

Livestock Activities

Livestock production activities for beef cows, hogs, dairy, and beef feeding were defined for each of the 10 consuming regions. National production activities were defined for hens and chickens, broilers and turkeys, and sheep and lambs. Each national activity withdraws feed (total digestible nutrients (TDN), protein, and roughage) from each consuming region in proportion to historical patterns.

Limits on the levels of the livestock activities are defined as constraints in each consuming region. The hog capacity and milk cow capacity for a consuming region are defined as the maximum, historical number of hogs and milk cows in annual inventory in that region between 1959 and 1968. The regional capacity constraints for beef cows and for fed beef are based on historical trends in annual inventory numbers of each region between 1959 and 1968.

The actual feed required per unit of production for each live animal, except for dairy and broilers and turkeys, is adjusted from the levels given in the USDA series on Livestock and Meat Statistics (USDA, Economic Research Service and Statistical Reporting Service, 1975) and other government sources. Feed requirements and milk production per dairy cow are estimated recursively to provide consistent projections of relations between feed input and milk output. Data from 1949 to 1969 are used in the following recursive system in each state:

$$F_t = a_0 + a_1T + e_t \tag{9.18}$$

$$M_t = b_0 + b_1F'_t + u_t \tag{9.19}$$

where F_t = total feed intake measured in TDN in year t; M_t = milk per cow in year t; and F'_t = predicted TDN required per cow in year t. The individual state projections are aggregated into consuming-region-level projections using 1963-65 dairy cow numbers as weights.

Projected feed required for broiler and turkey production is obtained from linear trends using an autoregressive least squares technique as described in Fuller and Martin (1961).

Cost Projection

For crop activities exogenous costs are categorized and estimated as labor, fertilizer, and other capital costs. Exogenous costs for livestock activities are labor and other capital costs. Labor costs are projected to 1980 by using an index developed by relating relative labor requirements for each commodity to changes in farm size and lagged relative capital requirements. Stoecker (1974) developed this index to project 1964 requirements to 1980 for the 10 consuming regions and commodities within the model. Using 1963-64 prices, fertilizer costs are estimated from the optimal fertilizer applications for each crop in each production area. Other capital costs include all other exogenous costs besides fertilizer and labor. Eyvindson (1970) developed cross-section estimates of exogenous costs required per activity unit for the year 1965. Using these estimates, Stoecker (1974) developed a set of time series equations for cost projections to 1980.

Transportation

A transportation system is provided for cattle, hogs, manufactured milk, oils, wheat, corn, oats, barley, feedgrains, oil meals, feeder calves, and yearling cattle. The central cities in each consuming region are used for defining the distances and estimating the transportation costs. Transportation costs are functions of distance and the mileage rate.

The Mathematical Model

The objective function for the national model (9.20) maximizes net aggregate producer profit. The mathematical model is described below. To simplify reading, area and regional subscripts and superscripts are dropped, and it is implied that the terms are expanded to have one set for each producing area or consuming region as is appropriate to the activity or imputed price vector. The objective function of the model is thus:

$$\text{Maximize } f(p, w, p_s, u, x, z_1, z_2, z_3, q_d, q_i, q_s) = p(d + Dp) + w'e + p_s'e_s - u'r - x'c - q'_d t_d - q_i' t_i - q_s' t_s \qquad (9.20)$$

subject to:

$$Dp - A_d x + z_1 - z_2 - T'_d q_d \leq -d \qquad (9.21)$$

$$-A_i x - A_c z_3 - T'_i q_i \leq -e \qquad (9.22)$$

$$-A_s x - z_1 + z_2 - z_3 - T'_s q_s \leq -e_s \qquad (9.23)$$

$$A'_j x \leq r \qquad (9.24)$$

$$A'_d p + A'_i w + A'_s p_s - A'_j u \leq c \qquad (9.25)$$

$$-p + p_s = 0 \qquad (9.26)$$

$$A'_c w - p_s \leq 0 \qquad (9.27)$$

$$T_d p \leq t_d \qquad (9.28)$$

$$T_i w \leq t_s \qquad (9.29)$$

$$T_s p_s \leq t_w \qquad (9.30)$$

$$p, w, p_s, u, x, z_1, z_2, z_3, q_d, q_i, q_s \geq 0 \qquad (9.31)$$

Constraint (9.21) states that the supply of desired commodities must be greater than or equal to the demand for desired commodities. Constraints (9.22) and (9.23) state that the supply of intermediate and substitutable commodities must be greater than or equal to the demand for intermediate and substitutable commodities, respectively. Constraint (9.24) states that there is a limited supply of primary resources and no more than this maximum can be used in production.

Constraint (9.25) can be rewritten as:

$$A'_d p + A'_i w \quad -A'_j u \leq c \tag{9.25a}$$

$$A'_x p_s \quad -A'_j u \leq c_s \tag{9.25b}$$

to see the requirement of marginal revenue being equal to or less than marginal cost plus rent of primary resources.

Equality constraints (9.26) are required because of the assumed perfect substitutability between corn, oats, and barley for food and feed. Constraints (9.28) through (9.30) are Samuelson's (1952) requirements for trade equilibrium stated in a slightly different form.

To expand the model (eqn. 9.20-9.30) to include producing areas, consuming regions, and primary commodities, these subscript and (or) superscripts are used: h = producing area = 1, 2, ..., 103; k = consuming region = 1, 2, ..., 10; d = desired commodity = 1, 2, ..., 14; i = intermediate commodity = 1, 2, ..., 5; s = substitutable commodity between intermediate and desired commodities = 1, 2, 3; j = primary commodity = 1, 2, ..., 11. In the model the terms are defined as: P^k, w^k, P^k_s, $u^{h \text{ or } k}$ = vectors of imputed prices for desired, intermediate, substitutable, and primary commodities respective, in region k or area h; $x^{h \text{ or } k}$ = vector of production activities in area h for crop production and in region k for livestock production; D = a matrix of demand slope coefficients with the vector of intercepts, d (this demand matrix is partitioned into submatrices for regional, regional-national, and national relationships); z^k_1 = transfer activity for food grains to feed grain markets; z^k_2 = transfer activity for feed grains to food grain markets; z^k_3 = transfer activity for converting feed grains into the units of TDN and protein by a conversion matrix, A_c, for livestock production; e^k and e^k_s = vectors of exogenous demands for intermediate and substitutable commodities, respectively; $r^{h \text{ or } k}$ = vector of primary resources in area h or region k; $A^{h \text{ or } k}_d$, $A^{h \text{ or } k}_i$, $A^{h \text{ or } k}_s$, $A^{h \text{ or } k}_j$ = matrix of technical coefficients relating primary resources and other inputs into intermediate and desired commodities through production or transfer activities x and z in area h or region k; $c^{h \text{ or } k}$ = vector of unit activity costs for intermediate and desired commodities in area h or region k; $q^{kk'}_d$, $q^{kk'}_i$, $q^{kk'}_s$ = vectors of interregional shipment levels of desired, intermediate, and substitutable commodities, respectively, from k to k' where $k \neq k'$; t_d, t_i, t_s = vectors of transportation costs for those desired, intermediate, and substitutable commodities, respectively, for which transportation is defined; $T^{kk'}_d$, $T^{kk'}_i$, $T^{kk'}_5$ = transportation matrices for the respective commodities.

POLICY ALTERNATIVES

The model, as developed in the previous section, is solved to obtain a competitive equilibrium for U.S. agriculture in 1980 in which commodity prices and outputs are determined simultaneously with certain restrictions on input use. The first competitive equilibrium solution is utilized as a base or reference point to compare with additional solutions obtained from the model after modification to reflect the imposition of a specific policy alternative. One policy alternative imposed on the model is a limit on the rate of nitrogen fertilizer application. The second policy alternative is the removal of four organochlorine insecticides for use on agricultural crops. The solutions obtained from the model for these alternatives are compared to the base solution and any differences between the policy solutions and the base solution are attributed to the policy.

The modifications of the model to reflect the policy alternatives will be developed in the following sections and then the analysis of the solutions will be presented.

Nitrogen Fertilizer Restraint Alternative

To look at the possible effects of a policy of fertilizer restraints, the coefficients in the cropping activities of the model are modified according to the following assumptions: for corn and sorghum (both grain and silage) farmers can apply fertilizer up to 110 pounds of elemental nitrogen; for cotton, 80 pounds; for wheat, oats, and barley, 55 pounds; and for soybeans, no nitrogen fertilizer is allowed.

Each cropping activity in the model is checked for conformance to these assumptions. If an activity is using less than or just equal to these fertilizer restraints, then no changes are made. If an activity is using more nitrogen than allowed, the amount is adjusted, new yields calculated using the Spillman function, and costs changed accordingly.

The amounts of phosphorus and potassium also are adjusted downward with nitrogen. This procedure is based on the supposition that farmers will apply nitrogen, phosphorus, and potassium fertilizers in recommended or correct ratios.

Insecticide Restriction Alternative

The Base Alternative allowed corn producers in the Midwest to use Adrin, Dieldrin, Chlordane, and Heptachlor to protect their crops from soil insects. The insecticide restriction is formulated by selecting substitutes for these four insecticides and then making appropriate adjustments for cost and yield changes. The substitutes selected are Thimet, Mocap, Dasanit, and Furadan. These materials are more expensive than the four insecticides to be removed from the market and equally effective except for two insect problems: the first year insect complex of wireworms and grubs in corn following a grass crop and cutworm damage to corn grown in lowland areas.

More than 90 percent of the agricultural use of Aldrin, Dieldrin, Chlordane, and Heptachlor is on corn acreages in 3 USDA regions: Corn Belt, Lake States, and Northern Plains (Delvo, 1974; Andrilenas, 1974). Because of this fact and the lack of reliable data for other areas, we have restricted our yield losses and cost increases to the states in those 3 USDA regions: Ohio, Indiana, Illinois, Iowa, Missouri, Michigan, Wisconsin, Minnesota, Kansas, Nebraska, North Dakota, and South Dakota.

The substitutes and their percentage share of the substitute mix are Thimet (40 percent), Mocap (5 percent), Dasanit (15 percent), and Furadan (40 percent). Using the 1971-72 price list and this mix of insecticides at the recommended rates, the increased cost per acre over the organochloride insecticides are determined for each of these USDA regions:

USDA region	Increased cost (1963-65 dollars)
Corn Belt	$2.243
Lake States	$0.377
Northern Plains	$0.265

These costs (in 1973-75 dollars) are the cost increases that are added to every corn and corn silage activity in the specified regions. In addition, there are other cost increases and yield losses that occur with different degrees of insect infestation. These cost and yield effects are calculated by the assumptions described below.

For the insect complex which attacks corn and corn silage during the first year following a meadow-type crop (wheat, nonlegume and legume hay, oats, and barley) the assumptions are (1) 20 percent of the first year corn will suffer a yield loss of 10 percent without any insecticide applied, and (2) the substitute chemicals are assumed to be 50 percent effective against the complex (i.e., the yield losses resulting from no insecticide application are reduced by half when the substitutes are used). To obtain the yield loss for each activity, the following equation is used:

$$\text{Yield loss due to first year complex (bu. or tons)} = Y_1 \times F \times F_A \times L_1 \times S \quad (9.32)$$

where Y_1 is the yield in the Base Alternative (bushels or tons); F is the percent of first year corn in activity; F_A is the percent of the first year corn affected = 20 percent; L_1 is the percent yield loss without insecticides = 10 percent; S is the percent effectiveness of substitutes = 50 percent. For example, if the present yield of corn in an activity was 100 bushels per acre and if 50 percent of the corn in this activity is first-year corn (two years of corn, one year of alfalfa hay), there would be a yield loss of 0.5 bushels per acre or a new yield of 99.5 bushels per year.

The cutworm and the low wetlands insect complex cause problems when the organochlorine insecticides are removed. The assumed percentage of wetland acres infested by this complex varies among the three USDA regions. These percentages of infestations are: Corn Belt, 16 percent; Lake States, 15 percent; and Northern Plains, 4 percent. The percentage of infestation on all cropland in a production area is found by using the percent of wetlands in that area and the infestation for the appropriate USDA region.

It is assumed that 25 percent of the acres infested will be replanted and that the remaining 75 percent will not be replanted. Those acres replanted will suffer from a yield loss because of timeliness and a cost increase because of replanting. The yield loss is determined by length of growing season and whether the land is irrigated. The assumed percentage losses (including the estimated need for new seed) are:

Area	Timeliness yield loss
North of Iowa	28%
Iowa and East	22%
South of Iowa	18%
West of Iowa	18%
Irrigated land	28%

The costs increase because of additional labor and machinery needed to replant. Each acre that is replanted is estimated to incur 10 percent more in machinery cost and to use 20 percent more labor. The equations used for the infested wetland acres are:

$$\text{Timeliness yield loss (in bu. or tons)} = Y_1 \times L_2 \times R \times W \times I \quad (9.33)$$

$$\text{Machinery-cost increase} = C_1 \times L_3 \times R \times W \times I \quad (9.34)$$

$$\text{Labor-hours increase} = L_1 \times L_4 \times R \times W \times I \quad (9.35)$$

where Y_1 is yield in the Base Alternative (bushels or tons); C_1 is machinery cost in the Base Alternative (1963-65 dollars); L_1 is labor hours in the Base Alternative; L_2 is percent yield loss in the appropriate USDA region for the production area in which the activity is defined; L_3 is percent increase in machinery cost = 10 percent; L_4 is percent increase in labor hours = 20 percent; R is percent of the infested wetlands that are replanted = 25 percent; W is percent of wetlands in the production area; and I is percent of wetlands acres that are infested in the appropriate USDA region.

For the 75 percent of infested wetlands not replanted, it is assumed that 75 percent of those acres are treated in a rescue operation and 25 percent are not "rescued." Those acres not rescued have a 25 percent yield loss. Rescued acres have an additional cost of $2.14 (1973-75 dollars) per acre for the insecticides and have a 15 percent yield loss in addition to the rescue cost. The substitute chemicals are 50 percent effective in combatting these losses. The equations used for the infested wetlands that are not replanted are:

Yield loss if not "rescued" (bu. or tons) $= Y_I \times L_5 \times R_N \times T_N \times W \times I \times S$ (9.36)

Yield loss if "rescued" (bu. or tons) $= Y_I \times L_6 \times R_N \times T \times W \times I \times S$ (9.37)

Additional cost due to rescue operation $= (\$2.14) \times R_N \times T \times W \times I$ (9.38)

where symbols are as previously defined and in addition L_5 = percent of yield loss if neither replanted nor "rescued" = 25 percent; L_6 = percent yield loss if not replanted but "rescued" = 15 percent; R_N = percent of infested wetlands not replanted = 75 percent; T_N = percent of infested wetlands neither replanted nor "rescued" = 25 percent; and T = percent of infested wetlands not replanted but "rescued" = 75 percent. Equations (9.32) and (9.38) are used to estimate the microeconomic effects of removing Aldrin, Dieldrin, Chlordane, and Heptachlor from the marketplace. These microeconomic effects are incorporated into the base model to form the Insecticide Restriction Alternative.

THE BASE ALTERNATIVE

To study the potential impacts of the two environmental policies upon U.S. agriculture in 1980, a point of reference is needed. The solution to the unrestricted model, referred to as the Base Alternative, is a competitive equilibrium for U.S. agriculture without governmental controls on agricultural practices. The Base Alternative does not include price supports, export controls, or set-aside programs. Some of the equilibrium commodity prices and quantities and regional land utilization results from the Base Alternative will be presented to provide an initial perspective before the comparison with the results from the policy alternatives.

Estimates of commodity prices (in 1963-65 dollars) and consumption for 1980 from the Base Alternative are shown in Table 9.2. In interpreting the levels of the prices, several points should be kept in mind. These prices do not incorporate fixed costs but they will still serve as a basis for comparison with price levels found in the policy alternatives. In other words, if fixed costs were included, the equilibrium prices generated in all of the alternatives would be somewhat higher. However, since fixed costs are not included in any of the alternatives, they are comparable in level of prices. Thus comparison of price results between the different solutions is relevant. A set of normal export demands, as explained earlier, are projected to 1980. At these export levels U.S. agriculture still has "excess supply capacity," as it did up through 1972. In contrast to the 1970-72 period, however, the model does not include supply control, price support, and international food-aid programs such as those in effect prior to 1973. The prices reported here and in later sections are under conditions of a free market and "normal" exports and are of a short-run nature (expressing conditions before farmers would shift resource use in response to price levels). These prices are in 1963-65 dollars and thus do not include the effects of inflation of the past few years, but relative comparisons among alternatives are the same as if they were in 1984 dollars.

Table 9.2. National equilibrium prices, total domestic consumption, and per capita consumption in the Base Alternative

Commodity	Price[a]	Domestic Consumption	Per Capita Consumption
	($/lb.)	(mil. lb.)	(lb.)
Cattle	0.27	44,901	196.1
Calves	0.19	186	0.8
Hogs	0.15	22,120	96.6
Fresh Milk	0.02	48,393	211.4
Manufactured milk	0.02	82,075	39.0
Oils	0.02	8,922	358.5
Sheep and Lambs	0.22	1,194	5.2
Eggs[b]	0.21	6,091	26.6
Poultry	0.09	16,999	74.2
Cotton	0.02	3,892	17.0
	($/bu.)	(mil. bu.)	(bu.)
Wheat	1.30	486	2.1
Corn	0.86	488	2.1
Oats	0.46	61	0.3
Barley	0.76	133	0.6

[a]Weighted average of regional prices with production as weights. Measured in 1963-65 dollars.

[b]Unit on eggs is dozen to give the appropriate column headings: $/doz., mil. doz., and doz.

In the Base Alternative, 21.5 percent of the available cropland is uncropped after all domestic and foreign demands have been met (Table 9.3). There is considerable regional variability in the percentage of available cropland which is uncropped. Almost all of the available cropland in the Corn Belt region is cropped, while one-third to one-half of available cropland is idled in the Appalachian, Southeast, Northern Plains, and Mountain regions.

Table 9.3. Estimated total available cropland, acreage used, and percent uncropped in 1980 in the Base Alternative

Region[a]	Cropland		
	Available	Used	Percent uncropped
	(thousand acres)		
Northeast	6,773	6,332	6.5
Appalachia	11,622	6,558	43.6
Southeast	11,222	7,245	35.4
Delta	11,517	10,721	6.9
Corn Belt	70,500	68,465	2.8
Lake States	27,147	21,860	19.5
Northern Plains	62,971	37,750	40.1
Southern Plains	31,692	27,736	12.5
Mountain	17,870	8,822	50.6
Pacific	9,785	9,373	4.2
United States	260,999	204,862	21.5

[a]See Figure 9.1 for map of the regions.

THE FERTILIZER RESTRICTION ALTERNATIVE

Average crop yields in the Fertilizer Restriction Alternative are lower than in the Base Alternative (Table 9.4). These lower average yields are principally the result of the lower crop yields because of the restriction on fertilizer application rates. There also is an indirect effect as the reduced yields require that more acres be cropped. The additional acres that are placed in production, as land and other agricultural resources are substituted for fertilizer, are of lower productivity. The result is that crop yields, as a national average, are significantly reduced.

The lower productivity of the U.S. agricultural sector following the fertilizer restriction raises production costs. Because the model incorporates a demand sector in which consumption of agricultural commodities respond to changes in price levels, i.e., quantities and prices are determined simultaneously in the model's solution, total output of agriculture is less under the fertilizer restriction. U.S. farm income will also be affected as the inelastic demand for agricultural commodities raises farm income.

Resource Substitutions

Comparison of the Base Alternative and the Fertilizer Restriction Alternative in Table 9.4 shows the aggregate changes in agricultural resource use. Nationally, total fertilizer use, and thus costs, decreases from $1734 million in the Base Alternative to $1246 million in the Fertilizer Restriction Alternative, even though there is an increase of 11 percent in the total acreage used for crop production. Because the decreases in fertilizer costs are offset by higher labor and capital cost (for other inputs), total input costs are approximately the same in both solutions.

Table 9.4. Comparison of estimated values for selected variables in the Base and Fertilizer Restriction Alternatives

Variable	Units	Base Alternative	Fertilizer Restriction	Fertilizer Res.[c] / Base Alternative
		Units		
Fertilizer cost	(million dollars)[a]	1,734	1,246	0.72
Labor cost	(million dollars)	1,527	1,588	1.04
Capital cost	(million dollars)	8,743	9,215	1.05
Total cost	(million dollars)	12,003	12,048	1.00
Per capita farm-level costs[b]	(dollars)[a]	111.26	112.84	1.01
Cropland utilization	(1,000 acres)	204,862	226,330	1.10
Corn yield	(bu./acre)	90	78	0.87
Corn acreage	(1,000 acres)	46,919	50,400	1.07
Wheat yield	(bu./acre)	37	31	0.84
Wheat acreage	(1,000 acres)	41,401	48,544	1.17

[a]1963-65 real dollars.

[b]Calculated from per capita consumption and regional price estimates.

[c]Ratio of Fertilizer Restriction value to the Base Alternative value.

The combined effect of lowering the productivity of the crop sector with the fertilizer restriction and bringing additional lands into production (which also use nitrogen) is shown in Table 9.4. The yield and acreage changes for corn and wheat are shown in Table 9.4 because the changes for these two crops are the largest of the crops included in the model.

There are significant regional differences as more of the available cropland is placed in production to meet foreign and domestic demands for agricultural commodities. As shown in Table 9.5, the percent of uncropped land declines from 21.5 to 13.3 when the fertilizer application rate is restricted. There is no change in the Corn Belt, the region with the most available cropland and fewest uncropped acres. The changes in the utilization of available cropland are particularly dramatic in the Southeast, Northern Plains, and Mountain regions. The percentage of the available land that is cropped increases markedly in the Fertilizer Restriction Alternative compared to the Base Alternative in these regions. Such changes will greatly increase the farm income in these regions compared to the rest of the U.S. agricultural sector.

Table 9.5. Total available cropland, acreage used, and percent idle for the Base and Fertilizer Restriction alternatives

Regions[a]	Land Available	Fertilizer Restriction Alternative		Base Alternative
		Used	Uncropped	Uncropped
	----(thousand acres)----		%	%
Northeast	6,773	6,481	4.3	6.5
Appalachia	11,622	6,581	43.4	43.6
Southeast	11,222	8,650	22.9	35.4
Delta	11,517	10,443	9.3	6.9
Corn Belt	70,400	68,465	2.8	2.8
Lake States	27,147	22,196	18.2	19.5
N. Plains	62,971	54,913	12.8	40.1
S. Plains	31,792	28,089	11.4	12.5
Mountain	17,870	11,108	37.8	50.6
Pacific	9,785	9,404	3.9	4.2
U.S.	260,998	226,330	13.3	21.5

[a]See Figure 9.1 for map of the regions.

Price and Output Changes

The restriction on the use of fertilizer in the Fertilizer Restriction Alternative alters the productivity of the crop sector so significantly that a substantially changed set of equilibrium commodity prices and quantities are obtained when the model is solved. Price and quantity results for selected commodities are shown in Table 9.6 for both alternatives. The reduced productivity of the crop sector and the resulting higher cost of agricultural products interact with the demand sector of the model to generate a new set of equilibrium prices and quantities. In general, equilibrium output levels in the Fertilizer Restriction Alternative are less than the output levels in the Base Alternative. There are some exceptions.

able 9.6. Comparison of estimated aggregate values for selected comodities in the Base and Fertilizer Restriction Alternatives

	Base Alternative		Fertilizer Restriction		Fertilizer Restriction Base[c]	
ommodity	Quantity[a]	Price[b]	Quantity[a]	Price[b]	Quantity[a]	Price[b]
eat	1,519	1.30	1,485	1.44	0.98	1.11
rn	4,226	0.86	3,950	0.96	0.93	1.12
ts	701	0.46	773	0.51	1.10	1.11
ttle	28,916	27.30	28,391	27.50	0.98	1.01
gs	218,896	14.90	215,004	15.70	0.98	1.05

[a]Wheat, corn, and oats (million bu.), cattle (1,000 head), hogs (1,000 live cwt.).

[b]Wheat, corn, and oats ($/bu.), cattle and hogs ($/cwt.). Prices are in 1963-65 dollars.

[c]Ratio of production for the Fertilizer Restriction divided by production for the Base Alternative.

The price and output levels for the livestock products are affected by the changes in the crop sector through changes in the price level of the feedstuffs. The higher-cost feedstuffs result in higher-cost livestock products and, hence, a slightly smaller consumption of livestock products in the Fertilizer Restriction Alternative. Because of the domestic price increase in beef and pork prices, net international imports of these products increase slightly above the equilibrium levels in the Base Alternative. In addition, within the crop sector the fertilizer restrictions differentially reduce the productivity of corn and grain sorghum more than oats and barley. The result is that corn and grain sorghum become slightly more expensive relative to oats and barley in livestock rations in the Fertilizer Restriction Alternative. This is the explanation for the relatively large increase in the output of oats shown in Table 9.6.

National and Regional Farm Income

As compared to the Base Alternative, the Fertilizer Restriction Alternative has higher prices, lower yields, less production, lower consumption levels, and more land in crops. Because demand is inelastic, the fertilizer limitation causes the value of agricultural production, in total, to increase. (The values in Tables 9.2, 9.4, and 9.6 are based on the 1963-65 average value of the dollar. Increased to current dollar values, they would be considerably higher but would still bear the same relative magnitudes within either alternative or between alternatives.)

All regions do not benefit uniformly from this national increase in the value of farm production. For example, the Southern Plains and Pacific States regions have the largest change in the value of wheat production (Table 9.7). The largest change of corn production is found in the Delta and Corn Belt regions. The changes in income from livestock are relatively uniform across regions, with those regions with the largest livestock sectors having the largest increases.

Table 9.7. Estimated value of production of selected commodities by region for 1980

	Wheat		Corn		Cattle and Calves		Hogs	
Regions	Base	Fertilizer Restriction	Base	Fertilizer Restriction	Base	Fertilizer Restriction	Base	Fertili Restric
	(million dollars)[a]							
Northeast	257	273	165	178	3,331	3,349	911	940
Appalachia	92	98	121	133	984	989	274	283
Southeast	61	65	102	115	972	977	267	276
Delta	349	375	494	555	384	387	104	108
Corn Belt	120	144	148	171	2,106	2,119	532	554
Lake States	88	91	83	92	1,044	1,050	262	273
North Plains	11	12	7	8	258	262	68	71
Southern Plains	491	534	27	30	765	770	209	216
Mountain	18	21	17	17	479	482	133	138
Pacific	448	522	76	75	1,318	1,323	492	509
United States	1,935	2,134	1,240	1,374	11,641	11,706	3,252	3,369

[a]1963-65 real dollars.

INSECTICIDE RESTRICTION ALTERNATIVES

Removal of Aldrin, Dieldrin, Chlordane, and Heptachlor is shown by the Insecticide Restriction Alternative to have few effects on U.S. agriculture. Evidently, the areas and proportion of corn incurring losses and increases in costs are not large enough to have major effects at the national level. Based on the model's solutions, additional equilibrium prices and consumption change only slightly because of the insecticide restriction (Table 9.8).

Per acre insecticide expenditures do increase with the insecticide restrictions (Table 9.9). However, total production costs do not change significantly. The cost of pesticides used increases by 13.4 percent in the Insecticide Restriction

Table 9.8. Comparison of aggregate national values for selected commodities in the Base and the Insecticide Restriction Alternatives

	Base Alternative		Insecticide Restriction Alternative	
Commodity	Quantity[a]	Shadow Price[b]	Quantity[a]	Shadow Price[b]
Wheat	1,519	1.30	1,519	1.30
Corn	4,226	.86	4,208	.87
Soybeans	1,701	3.00	1,703	2.99
Cattle	28,916	27.30	28,911	27.30
Hogs	219	14.90	219	14.90

[a]Wheat, corn, and oats (million bushels), cattle (1,000 head), hogs (1,000 live cwt.).

[b]Wheat, corn, and oats (dollars/bushels), cattle and hogs (dollars/cwt.).

Alternative compared to their use in the Base Alternative, but total production costs increased by only 0.6 percent. In the Corn Belt region, one of the three USDA regions assumed to be affected by the insecticide ban, pesticide costs increase from $69 million in the Base Alternative to $99 million in the Insecticide Restriction Alternative.

Table 9.9. Comparison of estimated rational values of selected variables for the Base and the Insecticide Restriction Alternatives

Variable	Unit	Base Alternative	Insecticide Restriction	Insecticide Restriction / Base Alternative[c]
		--------(units)--------		--------(ratio)--------
Pesticide cost	(million dollars)[a]	247	280	1.13
Fertilizer cost	(million dollars)	1,734	1,732	1.00
Labor cost	(million dollars)	1,527	1,534	1.00
Capital cost	(million dollars)	8,495	8,527	1.00
Total cost	(million dollars)	12,003	12,073	1.01
Per capita farm-level costs[b]	(dollars)	111.26	111.35	1.00
Cropland utilization	(1,000 acres)	204,862	205,514	1.00
Corn yield	(bu./acre)	90	89	.99
Corn acreage	(1,000 acres)	46,919	47,230	1.00
Wheat yield	(bu./acre)	37	37	1.00
Wheat acreage	(1,000 acres)	41,401	41,231	1.00

[a]1963-65 real dollars.

[b]Calculated from per capita consumption and regional price estimates.

[c]Ratio of Insecticide Restriction values to Base Alternative values.

At the national level, aggregate land use changes little under the insecticide limitation. Impacts of the insecticide limitation are more noticeable at the regional level. The relative advantage of corn production in the Corn Belt region declines and 337,000 acres shift to the Northeast, Appalachian, and Delta regions (Table 9.10). Wheat production shifts from the Delta States and Northeast regions to the Corn Belt and Northern Plains regions. National and regional production patterns change little as a result of the insecticide restriction.

able 9.10. Comparison of estimated acreages for selected crops in regions most affected by the insecticide removal

rop	Corn Belt		Northeast		Appalachian	
	Insecticide Restriction	Base	Insecticide Restriction	Base	Insecticide Restriction	Base
	-----------(thousand acres)-----------					
orn	26,647	26,984	713	527	2,874	2,746
heat	4,209	3,449	2,248	2,435	0	0
oybeans	30,029	30,414	623	623	2,719	2,579
oughages	7,726	7,648	3,182	3,182	2,589	2,586

SUMMARY

In summary, under the conditions of normal trends in exports and absence of government programs of supply control and (or) price support, either restriction on nitrogen or insecticide use could be applied with only slight increases in farm commodity prices and consumer food costs. Regional production patterns would be altered under both restrictions, but only the nitrogen restriction causes major changes.

As with all potential policies, there are certain tradeoffs that must be remembered. Under the nitrogen restriction more land is needed to meet domestic and export demands; these additional lands may be from the fragile land and marginal land areas. In aggregate, more labor and capital are needed to produce, handle, and transport agricultural commodities. In regions particularly dependent upon high nitrogen usage for crop production, income and unemployment would decline. Similar impacts would occur under the insecticide limitation, although interregional shifts in production would not be as great as under the nitrogen limitation.

REFERENCES

Andrileans, P. A. 1974. Farmer's Use of Pesticides in 1971 ... Quantities. Agricultural Economic Report 252. USDA, Economic Research Service, Washington, D.C.

Brandow, C. E. 1961. Interrelations among Demands for Farm Products and Implications for Control of Market Supply. Pennsylvania Agricultural Experiment Station Bulletin 680. Pennsylvania State University, College Park.

Brokken, R. F. 1965. Interregional Competition in Livestock and Crop Production in the United States: An Application of Spatial Linear Programming, volumes 1-4. Unpublished Ph.D. dissertation, Iowa State University, Ames.

Chen, C. C. 1975. Quadratic Programming Models of U.S. Agriculture in 1980: With Alternative Levels of Grain Exports. Unpublished Ph.D. dissertation, Iowa State University, Ames.

Delvo, H. W. 1974. Economic Impact of Discontinuing Aldrin Use in Corn Production. ERS-557. USDA, Economic Research Service, Washington, D.C.

Eyvindson, R. H. 1970. A Model of Interregional Competition in Agriculture Incorporating Consuming Regions, Producing Areas, Farm Size Groups, and Land Classes, volumes 1-5. Unpublished Ph.D. dissertation, Iowa State University, Ames.

Fuller, W. A., and J. E. Martin. 1964. The Effect of Autocorrelated Errors on the Statistical Estimation of Distributed Lag Models. *Journal of Farm Economics* 43:71-82.

Goldman, A. J., and A. W. Tucker. 1956. Theory of Linear Programming. In *Linear Inequalities and Related Systems*, edited by H. W. Kuhn and

A. W. Tucker. Animals of Mathematics Study 38. Princeton University Press, Princeton, N.J.

Hall, H. H. 1969. Efficiency in American Agriculture: An Application of Quadratic Programming. Unpublished Ph.D. dissertation, Iowa State University, Ames.

Hanson, M. A. 1961. A Duality Theorem in Nonlinear Programming With Nonlinear Constraints. *Australian Journal of Statistics* 3(2): 64-72.

Heady, E. O., H. C. Madsen, K. J. Nicol, and S. H. Hargrove. 1972. Agricultural and Water Policies and the Environment: An Analysis of National Alternatives in Natural Resource Use, Food Supply Capacity and Environmental Quality. CARD Report 40T. Center for Agricultural and Rural Development, Iowa State University, Ames.

Ibach, D. B., and J. R. Adams. 1968. Crop Yield Response to Fertilizer in the United States. USDA Special Bulletin 431. Washington, D.C.

Meister, A. D., C. C. Chen, E. O. Heady. 1978. Quadratic Programming Models Applied to Agricultural Policies. Iowa State University Press, Ames.

Olson, K. D., E. O. Heady, C. C. Chen, and A. D. Meister. 1977. Estimated Effects of Two Environmental Alternatives in Agriculture: A Quadratic Programming Analysis. CARD Report 70. Center for Agricultural and Rural Development, Iowa State University, Ames.

Plessner, Y. 1965. Quadratic Programming Competitive Equilibrium Models for the U.S. Agriculture Sector. Unpublished Ph.D. dissertation, Iowa State University, Ames.

Samuelson, P. A. 1952. Spatial Price Equilibrium and Linear Programming. *American Economic Review* 42:283-303.

Stoecker, A. L. 1974. A Quadratic Programming Model of United States Agriculture in 1980: Theory and Application. Unpublished Ph.D. dissertation, Iowa State University, Ames.

Tucker, A. W. 1956. Dual System of Homogeneous Linear Relations. In *Linear Inequalities and Related System*, edited by H. W. Kuhn and A. W. Tucker. Annals of Mathematics Study 38. Princeton University Press, Princeton, N.J.

U.S. Department of Agriculture, Economic Research Service and Statistical Reporting Service. 1975. Livestock and Meat Statistics. Supplement for 1974. Special Bulletin 543. Washington, D.C.

CHAPTER 10. A SEPARABLE PROGRAMMING ANALYSIS OF ALTERNATIVE INCOME AND SOIL CONSERVATION POLICIES FOR U.S. AGRICULTURE

by William G. Boggess and Earl O. Heady

THE OBJECTIVE of this study is to investigate the potential for reconciling two of the nation's goals for U.S. agriculture: increasing farm income and conserving agricultural land resources. A national interregional separable programming model of U.S. agriculture was formulated and employed in the analysis of alternative supply control and soil conservation policies designed to achieve these goals. The separable programming model allows incorporation of demand relationships so that equilibrium commodity prices and quantities are determined simultaneously and endogenously to the model. This model is presented in the following section and the results are discussed in the final section of this chapter.

THE MODEL

This model uses the same 105 producing areas and 28 market areas as explained in Figures 3.1 and 3.2, respectively. Each producing area has five land groups as summarized in Table 4.1. The model also uses the same endogenous and exogenous activities as in Chapter 3. Hence, livestock, fruits, and vegetables are determined exogenously. However, their feed and resource-demand quantities are all incorporated into the model.

The Demand Sector

The demand sector consists of three parts: (1) demands for endogenous commodities for which demand equations are specified, (2) demands for endogenous commodities with fixed demands, and (3) demands for exogenous commodities. Each of these demands incorporates net export, domestic consumption, and intermediate uses of the commodities.

Demands for the exogenous commodities (dry beans, dry peas, flaxseed, fruits and nuts, peanuts, potatoes, rice, rye, sugar cane, sweet potatoes, tobacco, sugar beets, miscellaneous other crops, poultry, turkeys, sheep and lambs, horses and mules, and miscellaneous other livestock) are obtained from USDA projections (Quance et al., 1977). Adjustments are made in the model's resource base for the quantities of land, water, nitrogen, and feedstuffs utilized in the production of these commodities.

Cotton is the only endogenous commodity with fixed demands. Cotton demands equal the sum of 1985 projected domestic uses and net exports. The model allocates land, water, and nitrogen to produce the specified quantity of cotton.

Domestic demands for the rest of the endogenous commodities (beef, pork, milk, wheat, feed grains, and oils) are determined within the model. Intermediate uses of wheat, feed grains, and oilmeals are endogenously determined within the livestock sector. Net exports of the commodities are fixed according to government projections for 1985 (Quance et al., 1977). The demand for each commodity is a function of its price, the prices of the five other commodities, consumer income, and the index of nonfood prices. Linear demand slopes were derived from Brandow's (1961) elasticities for these endogenous commodities of beef, pork, milk, wheat, feed grains, and oils using the following relationship:

$$\delta q(i)/\delta p(j) = e(i, j) * q(i)/p(j) \tag{10.1}$$

where e(i, j) is the percentage change in quantity demanded of the ith commodity due to a percentage change in the price of good j; q(i) and p(j) are the 1974-76 average quantities and prices; and $\delta q(i)/\delta p(j)$ is the linear slope showing the change in quantity demanded of good i due to a change in price of good j.

Demand relationships were specified for each of the 8 demand regions. Since the national demand matrix (Table 10.1) is consistent with a summation over individual consumer-demand equations, the regional demand matrices are derived from the national demand matrix as:

$$B'(i) = w(i) * D \tag{10.2}$$

where B'(i) is the 6×6 matrix of demand slopes for demand region i; w(i) is the proportion of total population in the ith demand region; and D is the 6×6 national demand matrix shown in Table 10.1. The regional demand intercepts are derived from the national demand intercepts by adjusting for expected regional differences in personal disposable income as follows:

$$d(i) = d + dI[pI(i) - pI(US)] \tag{10.3}$$

where d(i) is the ith region domestic-demand equation intercept; d is the national domestic-demand intercept; dI is a 6×1 vector relating changes in personal disposable income to the quantity demanded at the national level; and pI(i) is expected personal disposable income per capita for the ith demand region.

Table 10.2 lists the projected population and personal disposable income by demand regions for 1985. Series E population and per capita disposable income estimates were made by the Bureau of Economic Analysis (U.S. Water Resources Council, 1974).

Table 10.1. National farm-level demand for food use, 1985[a]

Commodity	Beef	Pork	Milk	Wheat	Feed Grains	Oils
Beef	-8,533.90	122.33	240.19	136.84	56.01	729.90
Pork	683.97	-2,183.10	105.31	59.96	149.29	2.24
Milk	352.74	121.25	-55,370.40	173.97	70.51	11,738.30
Wheat	433.37	149.42	804.70	-4,212.97	318.72	15.64
Feed grains	278.13	95.98	516.19	499.38	-10,429.50	10.05
Oils	36.35	12.53	2,469.70	18.60	7.66	-5,434.50

[a]Slope coefficients showing the effect of a one-unit change in the farm price of the commodity (at the head of a column) on the demand for the commodities (at the left).

Table 10.2. Projected population and personal disposable income by demand region for 1985[a]

Region	Population	Personal Disposable Income[b]	Proportion of U.S. Population
	(million)	(dollars)	
Northeast	60.482	7,963	0.2579
Southeast	28.963	6,133	0.1235
Lake States	31.027	7,668	0.1323
Corn Belt	37.921	6,881	0.1617
Delta States	11.773	5,472	0.0502
Northern Plains	13.766	6,699	0.0587
Southern Plains	19.418	6,388	0.0828
Pacific	31.167	7,647	0.1329
United States	234.517	7,157	1.0000

[a]Source: (U.S. Water Resources Council, 1974).

[b]Measured in 1974-1976 average dollars.

Land Base

The cropland base is built from the Conservation Needs Inventory (CNI), which reports acres of land by use and by agricultural capability class (USDA, Conservation Needs Inventory Committee, 1971). The CNI uses eight major capability classes, with classes II-VIII further subdivided to reflect the most severe hazard prevents the land from being available for unrestricted use. The subclasses reflect susceptibility to erosion (e), subsoil limitations (s), drainage problems (w), and climatic conditions preventing normal crop production (c).

The original CNI county acreages are aggregated, for dryland and irrigated uses, to the 105 producing areas by the twenty-nine capability classes to give the 5 land quality groups shown in Table 4.1. Irrigated acreages are specified separately for the seventeen western states (producing areas 48-105). Prior adjustment was made for land used by the exogenous crops. This adjustment was justified on the basis that these crops are generally the higher-value and location-specific crops, which have economic advantage in competition for land use.

Crop Production Sector

As in Chapter 8 the crop sector includes barley, corn, corn silage, cotton, legume hay, nonlegume hay, oats, sorghum, sorghum silage, soybeans, and wheat. Unique production activities are defined for each of 5 land quality groups in each of the 105 producing areas. Each crop management activity consists of a rotation, a tillage and conservation practice, and irrigated or dryland farming utilizing the nitrogen, land, and water resources available by land group. Crop production costs include machinery, labor, pesticide, and miscellaneous costs. These costs were developed using the United States Department of Agriculture (USDA) Firm Enterprise Data System (FEDS) crop budget generator.

Gross soil loss as specified in the model is the average annual tons of soil leaving the field. This measurement of soil loss does not represent the amount reaching the nation's waterways. As in earlier chapters, two separate procedures were used to determine gross soil loss per acre. For the areas east of the Rocky Mountains the Universal Soil-Loss Equation was used. For areas west of the Rocky Mountains, soil-loss coefficients for each management system were derived from a Soil Conservation Service questionnaire (Nicol and Meister, 1975).

Livestock Sector

The endogenous livestock sector includes beef cows, feeders, dairy, and hogs. Several alternative rations, reflecting optimum input ratios for alternative output levels and price ratios, are specified for each type of livestock. Endogenous selection of the least-cost ration then is allowed. Activity costs include all production costs except feed, which is endogenously determined. Nitrogen in livestock manure is a source of fertilizer to the crop sector.

Water Sector

The water sector defines water supplies in the western United States as shown in Figure 3.1. Dependable supplies of both surface and ground water are defined (Colette, 1976). Water prices are acreage-weighted, average reimbursable costs of the Bureau of Reclamation water projects. Water transportation activities are defined reflecting both natural and man-made flows.

Transportation Sector

Interregional interdependence is introduced through the transportation sector for the 8 market regions and their demands. Transportation routes are defined between all contiguous regions. Transportation costs are based on 1975 rail rates for grains and truck rates for livestock commodities using the mileage between the metropolitan centers of the market regions.

Mathematical Description of the Model

This study uses separable programming to approximate a nonlinear objective function with linear segments. The demand data developed by Stoecker (1974) is updated and used to specify the demand relationships. These demand relationships are converted to the demand formulation specified by Duloy and Norton (1975) by solving for prices as a function of quantities.

$$p = -D^{-1}d_0 + D^{-1}q \qquad (10.4)$$

$$= a + Bq$$

The objective function maximizes the area between the demand curve and the supply curve:

$$\text{Maximize } W = \int_0^{q} n \ldots \int_0^{q} 1(a + Bp)\, dq - c(q) \qquad (10.5)$$

where c(q) is an n × 1 vector of total cost functions and W is the sum of producer and consumer surpluses. Integration of equation 10.5 yields:

$$\text{Maximize } Z = q'(a + .5Bq) - c(q) \qquad (10.6)$$

Equation 10.6 is quadratic in q. Expansion of the term $.5q'Bq$ results in squared and cross-product terms. The cross-product terms are not separable since they are functions of two variables. This function is transformed to a separable function using a transformation outlined by Hadley (1964).

Once the objective function is transformed to separable form, grid and functional equations are specified for each variable. A separable programming routine is then used to incorporate the new objective function into the conventional linear programming framework.

Objective function. The objective function incorporating both the separable demand and the supply sides of the markets may be expressed mathematically as:

$$\begin{aligned} \text{Max. } Z = & \sum_n \sum_s A_{ns} Q_{ns} + \sum_n \sum_s \sum_r \sum_v B_{nsrv} SV_{nsrv} \\ & - \sum_i \sum_j \sum_k \sum_m X_{ijkm} XC_{ijkm} \\ & + \sum_n \sum_s \sum_q L_{nsq} LC_{nsq} - \sum_i W_i WC_i - \sum_i IB_i IC_i - \sum_n \sum_s \sum_t T_{nst} TC_{nst} \end{aligned} \qquad (10.7)$$

A definition of these terms and subscripts follows the mathematical expression of the constraints.

Model constraints. Constraints on the model include acreage of dry and irrigated land available for production of the endogenous crops, acre feet of water available for use by endogenous crops and livestock, commodity balance rows, a national cotton-demand equation, and the grid and constraint rows associated with the separable programming formulation of the demand equations. Mathematical expressions for each follow:

Dryland acreage by land class by producing area:

$$\sum_k \sum_m X_{ijkm} AD_{ijkm} \leq DA_{ij} \quad (10.8)$$

Irrigated land acreage by land class by producing area:

$$\sum_k \sum_m X_{ijkm} AI_{ijkm} \leq IA_{ij} \quad (10.9)$$

Water availability by producing area:

$$\sum_j \sum_k \sum_m \sum_u X_{ijkm} W_{ijkmu} CWU_{iu} + \sum_n \sum_p \sum_q L_{nsq} LWU_{nsq} LW_{nsi} \leq WS_i \quad (10.10)$$

Endogenous crop commodity balance rows by demand region:

$$\sum_i \sum_j \sum_k \sum_m X_{ijkm} W_{ijkmu} CY_{ijkmsu} - Q_{ns} - \sum_t T_{nst} \geq CD_{ns} \quad (10.11)$$

Endogenous livestock commodity balance rows by demand region:

$$LY_{nsq} L_{nsq} - Q_{ns} \geq LD_{ns} \quad (10.12)$$

Cotton demand at the national level:

$$\sum_i \sum_j \sum_k \sum_m X_{ijkm} W_{ijkmu} CY_{ijkmu} \geq DC \quad (10.13)$$

Subscripts in equations (10.7)-(10.13) refer to the following: n = 1, ..., 8 for the demand regions; s = 1, ..., 6 for the endogenous commodities; r = 1, ..., 6 for the endogenous commodities; v = 1, ..., 10 for the number of steps defined for each separable variable; i = 1, ..., 105 for the producing areas; j = 1, ..., 10 for the land quality classes; k = 1, ..., 330 for the crop rotations; m = 1, ..., 12 for the alternative conservation and tillage practices per rotation; q = 1, ..., 32 for the livestock rations; t = 1, ..., 76 for the transportation routes defined; and u = 1, ..., 10 for the possible irrigated crops.

Variables in equations (10.8)-(10.13) are defined as follows: A_{ns} is the linear demand coefficient for commodity s in region n; Q_{ns} is the number of units of commodity s marketed in demand region n; SV_{nsrv} is the separable variable step v, defined for each squared (s=r) and cross-product ($\neq$r) term for commodities marketed in region n; B_{nsrv} is the demand coefficient corresponding to step v of each of the squared and cross-product terms for commodities marketed in region n; X_{ijkm} is the level of rotation k using conservation-tillage method m on land class j in producing area i; XC_{ijkm} is the cost per acre of rotation k with conservation-tillage practice m in producing area i on land class j; L_{nsq} is the number of units of livestock activity s receiving ration q in demand region n;

LC_{nsq} is the cost per unit of livestock activity s receiving ration q in demand region n; W_i is the number of acre feet of water purchased in producing area i; WC_i is the cost per acre foot of water purchased in producing area i; IB_i is the acre feet of water transferred out of producing area i; IC_i is the cost differential on a per acre foot basis for water in producing area i; T_{nst} is the number of units of commodity s transported over route t from demand region n; TC_{nst} is the cost per unit of commodity s transported over route t from demand region n; AD_{ijkm} is the acres of dryland used per unit of rotation k using conservation-tillage method m on land class j in producing area i; DA_{ij} is the acres of dryland available for endogenous crops on land class j in producing area i; AI_{ijkm} is the acres of irrigated land used per unit of rotation k using conservation-tillage method m on land class j in producing area i; IA_{ij} is the acres of irrigated land available on land class j in producing area i; W_{ijkmu} is the rotation weight for crop u in rotation k using conservation-tillage method m on land class j in producing areai; CWU_{iu} is the acre feet per acre water-use coefficient for crop u in producing area i; LWU_{nsq} is the acre feet per unit water-use coefficient for livestock type s consuming ration q in market region n; LW_{nsi} is the proportion of livestock type s from demand region n falling in producing area i; WS_i is the acre feet of water available for use by the endogenous agricultural sector in producing area i; CY_{ijkmsu} is the per acre production of commodity s from crop u in rotation k using conservation-tillage system m on land class j in area i; CD_{ns} is the exogenously determined demand for commodity s in demand region n; LY_{nsq} is the yield of commodity s from livestock activity q in demand region n; LD_{ns} is the exogenously determined demand for livestock commodity s in demand region n; CY_{ijkmu} is the per acre production of cotton in rotation k using conservation-tillage practice m on land class j in producing area i; and DC is the exogenous national demand for cotton.

Alternatives Analyzed

Four alternatives of the model are specified and results analyzed to investigate the potential for reconciling two of the nation's goals for U.S. agriculture: increasing farm income and conserving agricultural land resources (Table 10.3). Alternative A is the base alternative for the study. This alternative is defined to represent the 1985 agricultural situation in the absence of specific governmental

Table 10.3. Description of the alternatives included in the study

Alternative	Description
A	Baseline trends and assumptions for 1985.
B	Soil erosion limited to tolerance (t) limits.
C	Ten percent of all cropland retired from production uniformly over all land groups.
D	Forty percent of cropland (the most fragile land in the grouping used) in land groups 3, 4, and 5 retired from production.[a]

[a]See Table 4.1 for definition of land groups.

measures to improve farm income or conserve soil. The solution to Alternative A is compared to the solutions from the other formulations of the model which incorporate governmental measures to achieve one or both goals. The differences between Alternative A and the other alternatives are identified as impacts of the policies analyzed.

Alternatives B, C, and D each take a different approach towards achieving the goals of increasing farm income and conserving soil. Alternative B is formulated with the principal objective of conserving soil without special concern for increasing farm income, Alternative C with the principal objective of increasing net farm income without special concern for soil conservation, and Alternative D with the dual objectives of increasing farm income and conserving soil. Consequently, the alternatives have quite different impacts on the degree of soil conservation achieved and the increase in income generated.

Alternatives B, C, and D each incorporate the baseline levels of production possibilities, population, per capita consumption, exports, etc. Alternative B differs from Alternative A only in that producers are required to limit gross soil erosion to established soil tolerance levels. These soil-loss tolerance levels range from 1 to 5 tons per acre year, depending upon soil properties, topography, and prior erosion. Producers will be able to choose any combination of crop sequence, conservation practice, and tillage practice that results in an estimated gross soil loss less than the tolerance (t) level. Alternative C differs from Alternative A only in that 10 percent of all cropland in all land groups is retired from production uniformly over the country. The principal objective is to reduce production and improve farm prices and income relative to Alternative A. This is the main supply control mechanism used by the government over the last 50 years. Level and productive land is retired from production along with the fragile and erosive land. With Alternative D, the only difference is the 40 percent of the cropland in land quality groups 3, 4, and 5 (the more fragile and erosive lands) of the model is retired from production in an attempt to simultaneously restrict production and reduce soil erosion. In Alternatives C and D the landowners receive a government payment to take the land out of production. This retired land is completely removed from production; neither substitution of other crops nor grazing by livestock is allowed. Alternative D differs from Alternative C in that only land in the more erosive and less productive land groups 3, 4, and 5 is retired, whereas in Alternative C the retired land is distributed uniformly over all lands in each land group.

RESULTS

In each of the three income and soil conservation policy alternatives analyzed in this study, net farm income is increased because of the interaction between rising production costs and inelastic commodity demands (see Table 10.4). The levels of soil erosion vary markedly between the policies. In Alternative B, the soil conservation alternative, both the national gross soil erosion and average per acre gross soil erosion are reduced to about one-third of the levels in Alternative

A. Total gross soil erosion is reduced only slightly in Alternative C using the 10 percent homogeneous land retirement policy. Alternative D, however, with the 40 percent retirement policy for lands most subject to erosion, not only results in a 15 percent increase in net farm income, but also achieves a 16 percent reduction in total gross soil erosion compared to Alternative A.

Table 10.4. Values for major variables at the national level under each alternative

	Units	Model Alternatives A	B	C	D
		(units)			
Total net farm income	million dollars	29,206	31,675	33,660	33,483
Soil erosion					
Total erosion	million tons	3,099.9	996.5	2,962.2	2,612.3
Per acre erosion	tons	8.7	2.7	8.9	8.0
Land utilized	thousand acres	377,053	377,758	353,622	346,734
Land retired	thousand acres	0	0	37,504	45,182
Land retirement payments	million dollars	0	0	2,057	1,556

The land retirement payments required under Alternatives C and D are significantly different (Table 10.5). Estimated government land retirement payments to farmers are .5 billion dollars less in Alternative D than in Alternative C, even though 20 percent more land is taken out of production in Alternative D. However, the same price-support levels and practically the same net farm income is attained as under the conventional supply control program, Alternative C. Net farm income under the soil conservation system of supply control (D) was less than 1.0 percent smaller than net farm income under the conventional supply control program.

Resource Use Changes

Imposing soil erosion control limitations on agricultural land affects the use of the other inputs employed in agriculture differently than land retirement programs do. The indices in Table 10.6 relating input use in Alternatives B, C, and D to Alternative A reflect the substantial increases in pesticides and fertilizer required in the crop production sector, especially since output is lower under Alternatives B, C, and D. The use of water in agriculture also increases.

This substitution of other agricultural inputs for land, especially in Alternatives C and D, is an important result obtained from the model. The larger income generated in Alternatives C and D results from increased prices associated with a shift in aggregate product supply functions. The actual decreases in production in C and D compared to A are relatively moderate for two reasons. The first is the inelastic demand for agricultural commodities. The second is that purchased

Table 10.5. Estimates of net farm income for each of the four alternatives

Item	Model Alternatives			
	A	B	C	D
	------------------(million dollars)--------------			
Endogenous cash receipts	71,661	76,723	82,709	81,341
Exogenous cash receipts[a]	27,046	27,046	27,046	27,046
Total cash receipts	97,707	103,769	109,744	108,387
Endogenous production expenses[b]	48,780	51,373	57,431	55,749
Exogenous production expenses[c]	29,892	29,892	29,892	29,892
Total production expenses	78,672	81,265	27,323	85,641
Net farm returns	20,035	22,504	22,432	22,746
Nonmoney income and inventory change[d]	9,171	9,171	9,171	9,171
Income from government payments[e]	0	0	2,057	1,556
Total net farm income	29,206	31,675	33,660	33,483

[a]1977 cash receipts for exogenous commodities reported in 1975 dollars.

[b]Expenses determined endogenous to the model.

[c]Other production expenses including $9,920 million for expenses related to the endogenous commodities and $19,972 million pertaining to exogenous commodities.

[d]1977 level of nonmoney income reported in 1975 dollars.

[e]Total payments equal acres retired multiplied by the per acre payments. The per acre payments equal the average of the land shadow prices in Alternative A and the corresponding land retirement alternative.

Table 10.6. Total input usage of purchased inputs and land and water resources in producing the endogenous commodities of the model

Item	Model Alternatives			
	A	B	C	D
Inputs (indices, A=100)				
Pesticides	100	123	121	111
Nitrogen fertilizer	100	116	126	117
Other fertilizer	100	109	105	102
Labor	100	104	102	101
Machinery	100	108	106	104
Total	100	107	105	103
Water used (acre ft)	66,787	68,430	67,883	68,662
Land (thousand acres):				
Cropped or retired	377,053	377,758	391,126	391,916
Unused	17,283	16,577	3,210	2,419

inputs (labor, chemicals, and machinery) substitute for land in the production process. For example, total purchased input use in C and D increases approximately 4 percent relative to A. The largest increases occur in chemical inputs (pesticides and fertilizers). Alternative D not only results in significantly less soil erosion than Alternative C but also in significantly less use of chemical inputs.

The increases in the use of pesticides and fertilizer in Alternative B reflects the consequences of changing the cropping methods to those practices that are more conserving of the soil (Table 10.7). The reduction in total gross soil loss with

Table 10.7. Percentage of acreage employing conservation and tillage practices for the four model alternatives

Practices	Model Alternatives			
	A	B	C	D
	--------------(percentage)-------------			
Tillage practices:				
Conventional tillage:				
residue removed	13	13	14	13
residue left	52	42	54	54
Reduced tillage	35	45	32	33
Conservation practices:				
Straight row	97	52	96	97
Contouring	2	29	2	2
Strip cropping	1	6	1	1
Terracing	1	13	1	1

Alternative B is achieved by reducing the row-crop intensity; increasing the use of contouring, strip cropping, and terracing; and shifting from conventional tillage to reduced-tillage practices. However, these changes in cropping methods were accompanied by increased use of chemical pesticides and fertilizers.

Price and Output Adjustments

The farm-level prices and production levels estimated in the four alternatives are presented in Table 10.8. The price and quantity changes from the base alternative obtained under the soil-loss tolerance limitation alternative are the results of the interaction between the higher production costs and the inelastic commodity demands. The higher production costs are the result of changes in farming methods and shifts in cropping patterns forced by the soil erosion limitation. These higher production costs raise the cost of the commodities to consumers. The interaction of these higher costs with the demand relationships defined in the model produces equilibrium prices and quantities that account for the demand response to higher prices, i.e., the higher the price of a commodity, the less that commodity is demanded. Farm-level commodity prices increase approximately 8 percent in Alternative B over the prices in Alternative A. Beef, pork, milk, and cotton prices increase by less than 8 percent while wheat, feed grains, silages, and hay prices increase by more than 8 percent. These price changes are accompanied by relatively small changes in output.

The price and quantity changes with Alternatives C and D result from the removal of over 37 million and 45 million acres of cropland, respectively. Prices increase an average of approximately 20 percent under the land retirement programs in both Alternatives C and D compared to prices under Alternative A. Individually, wheat, feed grains, silages, and hay prices increase by more than 20 percent in both solutions C and D. Beef, pork, milk, and cotton prices increase by less than 20 percent in both Alternatives C and D compared to these products' prices under Alternative A. These price changes from the land retirement

Table 10.8. Comparison of national average farm-level prices and national production of selected commodities[a]

Commodity	Unit	Model Alternatives A	B	C	D
Feed grains:					
Production	bu.	9,359,576	9,134,946	8,657,608	8,741,844
Price	$/bu.	1.92	2.10	2.50	2.43
Wheat:					
Production	bu.	2,281,708	2,233,792	2,076,354	2,115,143
Price	$/bu.	2.96	3.20	4.06	3.82
Beef:					
Production	cwt.	432,367	425,449	410,749	413,775
Price	$/cwt.	47.81	49.05	51.63	51.20
Pork:					
Production	cwt.	228,457	224,345	212,465	214,064
Price	$/cwt.	25.97	27.09	30.38	29.76

[a]The production and price estimates for Alternatives B, C, and D are adjusted based on empirical estimates of the actual elasticities of supply and demand to account for the fixed net export levels in the model. The procedure used to adjust the values is documented in Boggess (1979).

alternatives are larger than found with Alternative B and are also accompanied by correspondingly larger output decreases than with Alternative B. Price increases as large or larger than under C could have been attained under Alternative D. We simply made an estimate of the amount of land to be retired under D to give the same price levels as for C. We were so close that we decided not to undergo the computer expense of searching for the precise level of land to be retired under D in order to give the same prices as under C. Had we done so, net farm income under Alternative D would have slightly exceeded that under Alternative C.

Regional Production Practices

Changes in national production and yields obviously are important indicators of the impacts of the alternative policies. However, the relatively moderate changes at the national level may mask some important regional shifts in production that occur in response to changes in regional comparative advantages. Tables 10.9-10.12 report the acreages of the major crops by demand regions for each of the four model alternatives.

The regional distribution of total cropland used varies only slightly between alternatives, primarily because 95 percent or more of the cropland available is used in each of the alternatives. Thus the regional distribution of cropland used very nearly represents the regional distribution of cropland available. However, as the supply of available cropland declines, acres of summer-fallow decline from 20.8 million acres in Alternative A to 8.6 million acres in Alternative D. Thus the difference in acres of cropland actually cropped in the two alternatives is less than 20 million acres; despite the fact that over 45 million acres are retired in Alternative D.

Table 10.9. Endogenous cropland use for crops by demand regions for the baseline alternative (Alternative A)

Demand Region	Feed Grains	Wheat	Cotton	Soybeans	Roughages	Summerfallow	Total Used	Percent of U.S. Total
	(million acres)							
Northeast	6.5	2.4	0.0	0.2	2.8	0.0	11.9	3.3
Southeast	3.2	9.7	0.0	4.6	0.8	0.0	18.3	5.1
Lake States	28.4	15.9	0.0	9.3	5.7	0.0	66.5	18.6
Corn Belt	32.6	0.4	0.0	35.4	3.1	0.0	71.5	20.0
Delta States	7.0	2.7	4.8	7.3	3.2	0.0	25.0	7.0
Northern Plains	18.8	19.1	0.2	10.9	17.8	14.1	80.9	22.6
Southern Plains	30.5	5.1	1.7	5.7	16.6	1.7	61.3	17.1
Pacific	2.4	11.7	0.9	0.0	2.4	5.0	22.4	6.3
United States	129.5	67.0	7.6	73.4	52.2	20.8	357.8	100.0

Table 10.10. Endogenous cropland use for crops by demand regions for the soil loss alternative (Alternative B)

Demand Region	Feed Grains	Wheat	Cotton	Soybeans	Roughages	Summer-Fallow	Total Used	Percent of U.S. Total
	(million acres)							
Northeast	5.1	2.4	0.0	0.9	3.2	0.0	11.6	3.2
Southeast	9.5	3.9	0.0	2.6	2.1	0.0	18.1	5.0
Lake States	30.7	14.1	0.0	19.1	4.5	0.0	68.4	19.1
Corn Belt	29.7	10.0	0.0	28.1	3.2	0.0	71.0	19.8
Delta States	7.8	3.7	3.7	6.0	3.1	0.0	24.3	6.8
Northern Plains	22.9	17.8	0.3	16.0	13.7	10.1	80.8	22.5
Southern Plains	25.2	6.4	0.9	6.3	21.4	1.8	62.0	17.8
Pacific	2.7	12.0	2.0	0.0	3.7	1.9	22.3	6.2
United States	133.5	70.3	7.0	79.0	54.8	13.8	358.5	100.0

Table 10.11. Endogenous cropland use for crops by demand regions for the 10 percent land retirement alternative (Alternative C)

Demand Region	Feed Grains	Wheat	Cotton	Soybeans	Roughages	Summerfallow	Total Used	Percent of U.S. Total
	(million acres)							
Northeast	6.0	2.2	0.0	0.2	2.7	0.0	11.1	3.3
Southeast	0.9	12.1	0.0	3.1	0.3	0.0	16.5	4.9
Lake States	26.9	18.6	0.0	13.4	3.5	0.0	62.4	18.7
Corn Belt	31.3	0.9	0.0	30.9	2.7	0.0	65.8	19.7
Delta States	4.6	0.6	5.9	9.8	2.3	0.0	23.2	6.9
Northern Plains	19.7	21.1	0.1	15.2	11.5	9.1	76.6	23.0
Southern Plains	25.1	4.1	0.8	3.9	23.9	0.0	57.8	17.3
Pacific	0.6	13.1	0.7	0.0	6.2	0.1	20.7	6.2
United States	115.2	72.7	7.4	76.6	53.2	9.2	334.2	100.0

able 10.12. Endogenous cropland use for crops by demand regions for the 40 percent conservation land retirement alternative (Alternative D)

emand Region	Feed Grains	Wheat	Cotton	Soybeans	Rough-ages	Summer-fallow	Total Used	Percent of U.S. Total
	(million acres)							
ortheast	6.0	2.1	0.0	0.1	2.7	0.0	10.9	3.3
outheast	0.9	12.9	0.0	2.8	0.3	0.0	16.9	5.2
ake States	27.6	18.7	0.0	13.5	3.2	0.0	65.0	19.9
orn Belt	31.1	0.3	0.0	32.6	2.7	0.0	66.7	20.4
elta States	4.8	1.0	6.2	9.9	2.1	0.0	24.1	7.4
orthern Plains	17.9	18.8	0.1	12.6	11.5	8.5	69.4	21.2
outhern Plains	24.9	3.3	0.8	3.5	22.3	0.0	54.8	16.7
acific	0.9	12.9	0.4	0.0	5.1	0.1	19.4	5.9
nited States	114.0	70.1	7.6	75.1	50.1	8.6	327.2	100.0

The shift from summer-fallow to crops is only one of the changes that occurs in response to changes in comparative advantages between crops within a region and to changes in interregional comparative advantages. Under the soil-loss restrictions of Alternative B, acreages of cropland shift from the more erosive areas in the eastern United States to the less erosive western United States. Acres of crops increase by nearly 10 million acres in the Lake States, Northern Plains, Southern Plains, and Pacific regions and decline by nearly 2 million acres in the eastern United States. These regional shifts in acreage reflect changes in interregional comparative advantages brought about by the inclusion of soil erosion concerns in the producers' decision framework. Acres of the more erosive crops, soybeans and cotton, decrease by 11.0 million acres in the eastern United States while acres of less erosive crops, wheat and roughages, increase by 6.5 million acres. Total feed-grain acreage increases in both the eastern and western areas, but there is a shift from corn and sorghum to the less erosive small grains barley and oats especially in the more erosive areas.

In general, total acres used decline in all regions under both of the land retirement policies (Alternatives C and D). However, if summer-fallow acres are excluded, acres in crops increase in the Pacific region in both alternatives and in the Northern Plains in Alternative C. In these areas, and to a lesser extent in the Southern Plains, producers merely trade off summer-fallow acres for acres retired.

The regional distribution of class 3, 4, and 5 lands is quite similar to the regional distribution of all cropland, as evidenced by the similar total cropland use values in Alternatives C and D. The Lake States, Northern Plains, and Southern Plains regions are the only areas where total cropland used varies by more than 2 million acres between the two solutions. The increase in acres cropped in the Lake States in Alternative D compared to acres cropped in Alternative C indicates that a higher-than-average percentage of the region's cropland falls in land Classes 1 and 2. Similarly, the decline in acres cropped in the Northern and Southern Plains regions in Alternative D relative to those cropped in Alternative C indicates that a disproportionate share of the cropland in these regions falls in land Classes 3, 4, and 5.

On a national basis, the acreage of feed grains declines while the acreages of wheat and soybeans increase in Alternatives C and D relative to feed grains and wheat and soybeans acreages in Alternative A. These differences in total acreages occur in spite of the fact that total production of all three commodities declines in both Alternatives C and D. In Alternative C, feed-grain production declines by only 4 percent while acreage declines by 11 percent; the difference is made up by an increase in yields. Production of wheat and soybeans, on the other hand, declines by less than 1 percent while acreage actually increases by 9 and 4 percent, respectively, due to lower average yields for wheat and soybeans in Alternative D. These results suggest that the demand pressures on land created by the land retirement policies affect the comparative advantage both within and between regions in a manner which concentrates feed-grain production on the more productive land. The regional distribution of feed-grain acreage varies only slightly between Alternatives A and C, suggesting that the primary adjustments are made within regions by concentrating feed-grain production on Class 1 and 2 lands or by using irrigated land for feed-grain production. The implication of this adjustment is that feed grains exhibit a greater response to land productivity than do wheat and soybeans. Thus, since land is the most constraining factor of production, it is employed in a manner maximizing the comparative advantages in its use.

Regional Shifts in Net Farm Income

The differential impacts of the policies on regional net farm returns from the endogenous crops is an important measure of the contribution of these policies toward the goal of raising farm income. Regional net farm returns from the endogenous crops are reported in Table 10.13. The soil conservation policy of Alternative B has the smallest impact of the alternatives analyzed. With the exception of the Northeast, no region's share of net returns from the endogenous crops varies by more than 2 percent of the base. A shifting of endogenous crops

Table 10.13. Estimates of regional net farm returns from the endogenous commodities for each of the alternatives

	Alternatives							
	A		B		C		D	
Demand Region	Value[a]	Percent[b]	Value	Percent	Value	Percent	Value	Percent
Northeast	418	3.2	111	0.7	147	1.0	155	1.0
Southeast	571	4.4	838	5.4	915	6.0	807	5.1
Lake States	1,277	9.9	1,354	8.8	1,349	8.8	1,416	9.0
Corn Belt	2,979	23.0	3,810	24.6	5,435	35.4	5,586	35.6
Delta States	325	2.5	478	3.1	831	5.4	862	5.5
Northern Plains	2,413	18.6	2,797	18.1	2,983	19.4	2,863	18.3
Southern Plains	4,428	34.2	5,556	35.9	3,533	23.0	3,799	24.2
Pacific	550	4.2	515	3.3	166	1.1	186	1.2
United States	1,2961	100.0	15,459	100.0	15,359	100.0	15,673	100.0

[a]Values are in millions of 1975 dollars.

[b]Percent of U.S. total by region.

out of the Northeast region, in conjunction with the increased cost of production incurred in response to the soil-loss restrictions, results in the relatively large decline in net returns in the Northeast.

Alternatives C and D have quite similar effects on the regional distribution of net farm returns. Regional shares increase in the Southeast, Corn Belt, and Delta States regions and decline in the Northeast, Lake States, Southern Plains, and Pacific regions under both C and D. The largest increase occurs in the Corn Belt, where the share of national net farm returns increases by over 12 percent in Alternatives C and D relative to that under A. The largest decline occurs in the Southern Plains, where that region's share of returns declines by over 10 percent in both Alternatives C and D relative to its returns in Alternative A.

SUMMARY

This study indicates that a conservation-oriented land retirement program can be designed that will both increase net farm income and decrease gross soil erosion. The comparison of the two land retirement policies in this analysis demonstrates that a land retirement program for land susceptible to erosion can raise net farm income as effectively as a general land retirement program that places no special emphasis on soil conservation.

The conservation-oriented land retirement program in this study significantly reduces soil erosion and requires less pesticides and fertilizer as substitutes for land than the general land retirement policy. In addition, the burden of the conservation-oriented land retirement policy on public funds is less than with the general land retirement policy.

These results provide useful and valuable information to government policymakers. The study demonstrates the use and value of a national programming model in designing policies for U.S. agriculture.

REFERENCES

Boggess, William G. 1979. Development and Application of an Interregional Separable Programming Model for United States Agriculture in 1985. Unpublished Ph.D. dissertation, Iowa State University, Ames.

Brandow, G. E. 1961. Interrelations among Demands for Farm Products and Implications for Control of Market Supply. Pennsylvania Agricultural Experiment Station Bulletin 680. Pennsylvania State University, State College.

Colette, W. Arden. 1976. The Conceptualization and Quantification of Water Supply Sector for a National Agricultural Analysis Model Involving Water Resources. CARD (Center for Agricultural and Rural Development) Miscellaneous Report. Iowa State University, Ames.

Duloy, John H., and Roger O. Norton. 1975. Prices and Incomes in Linear Programming Models. *American Journal of Agricultural Economics* 57:591-600.

Hadley, G. 1964. Nonlinear and Dynamic Programming. Addison-Wesley Publishing Company, Inc., Reading, Massachusetts.

Meister, Anton D., and Kenneth J. Nicol. 1975. A Documentation of the National Water Assessment Model of Regional Agricultural Production, Land and Water Use, and Environmental Interaction. CARD Miscellaneous Report. Center for Agricultural and Rural Development. Iowa State University, Ames.

Quance, Leroy, Allen Smith, and Levi Powell. 1977. Scenario for Change in Food and Agriculture. No. 1 in the Regional Planning Series. USDA, Economic Research Service and Cooperative State Research Service, Washington, D.C.

Stoecker, Arthur Louis. 1974. A Quadratic Programming Model of United States Agriculture in 1980: Theory and Application. Unpublished Ph.D. dissertation, Iowa State University, Ames.

U.S. Department of Agriculture, Conservation Needs Inventory Committee. 1971. National Inventory of Soil and Water Conservation Needs 1967. USDA Statistical Bulletin No. 461. Washington, D.C.

U.S. Water Resources Council. 1970. Water Resource Regions and Subareas for the National Assessment of Water and Related Land Resources. U.S. Government Printing Office, Washington, D.C.

___. 1974. 1972 OBERS Projections, vols. 1-7. U.S. Government Printing Office, Washington, D.C.

Wischmeier, W. H., and D. D. Smith. 1965. Predicting Rainfall-Erosion Losses from Cropland East of the Rocky Mountains. USDA Agricultural Handbook No. 282. Washington, D.C.

CHAPTER 11. A REGIONAL-NATIONAL RECURSIVE MODEL FOR THE STATE OF IOWA

by James A. Langley, Burton C. English, and Earl O. Heady

THE MODEL described and the test results presented in this chapter are part of a cooperative effort between the Center for Agricultural and Rural Development (CARD) and the International Institute for Applied Systems Analysis (IIASA) in Laxenburg, Austria. IIASA is an international research institute with a major research effort addressing food and agricultural problems. This cooperative effort is concerned with examining the important relationships between agricultural production technologies, resource use, and the environment that will affect the stability and sustainability of the food and agricultural system in the long run. In order to carry out this investigation, a general modeling system is developed incorporating the regional nature of resource inputs and environmental impacts of agricultural production for the state of Iowa within a national framework.

OBJECTIVES OF THE STUDY

The specific objective of the CARD-IIASA project is to evaluate both the regional and national impacts of selected legislative policies of the state of Iowa aimed at controlling pollution resulting from soil erosion and sedimentation. A regional programming model of Iowa is linked recursively with a U.S. national econometric simulation model to determine the effects of such legislation upon production patterns; commodity supply, demand, and price; and other economic variables. Focus is upon the state of Iowa with necessary attention given to Iowa's relationship within the agricultural economy of the United States. An additional incentive for this study is the recent interest in "targeting" relative to soil erosion and environmental problems. Under current concepts there are plans to target areas which are most erosive and thus need to receive more intensive public investments and technical assistance in controlling erosion. Among others, regions well adapted to targeting are western Tennessee, western Iowa, and the Palouse area of the Northwest.

In the first part of this chapter, a general overview of the model is presented to give the reader a better understanding of the framework for the analysis. The next part presents an overview of the components of the regional-national recursive programming model. This section discusses the formulation of both the Iowa programming model and the national econometric simulation model and the method of linking the components together. The results of selected test runs of the model are then presented. These test solutions indicate how the model estimates the impacts of soil loss control policies upon agricultural production in Iowa. Finally, a summary and conclusions are given in the last section of the chapter.

THE MODELING FRAMEWORK

The regional-national recursive model developed consists of three main components: a regional linear programming model for the state of Iowa, a national econometric simulation model for the United States, and a linkage procedure which transfers information between the programming and econometric models. The purpose of the linear programming component is to determine crop production and input use occurring solely within the state of Iowa. The national econometric simulation component estimates resource use and commodity output originating in the United States excluding Iowa. This section of the paper presents an overview of the regional-national recursive system, followed by a more detailed specification of the programming, econometric, and linkage components.

This regional-national modeling system benefits from the integration of information on the spatial pattern of regional supply, resource use, and technical structure of production (generated by the regional programming model) with the detailed information on market structure and prices of commodities and inputs (generated by the national econometric simulation model). The potential linkages between the programming and econometric simulation components that are considered in formulating the regional-national recursive system for agricultural policy analysis are illustrated in Figure 11.1. It is assumed that the impacts of any policy program can be translated into changes in costs of production and yields (e.g., nitrogen fertilizer) or outputs (e.g., soil loss) of the production process. Also, policy impacts upon relative price relationships may affect farmers' decisions to purchase certain inputs or to produce certain crops. Production costs and yields are adjusted (C) and used to determine the profitability of production (D). Relative price relationships among the endogenous crops are used to determine the range of regional production response through a flexibility constraint formulation (E), and to adjust the coefficients in the objective function of the regional linear programming component (F). The linear programming model then determines regional production (G) and input factor demands (H). Soil loss is determined as a function of regional input factor and land use (L). The supply and factor demand subsequently determine the prices of commodities (I) and input factors (J). These prices are then used to determine production costs, yields, and net profits for the next time period (C) and resource constraint adjustments (K). The entire process is repeated until the predetermined number of simulations are completed.

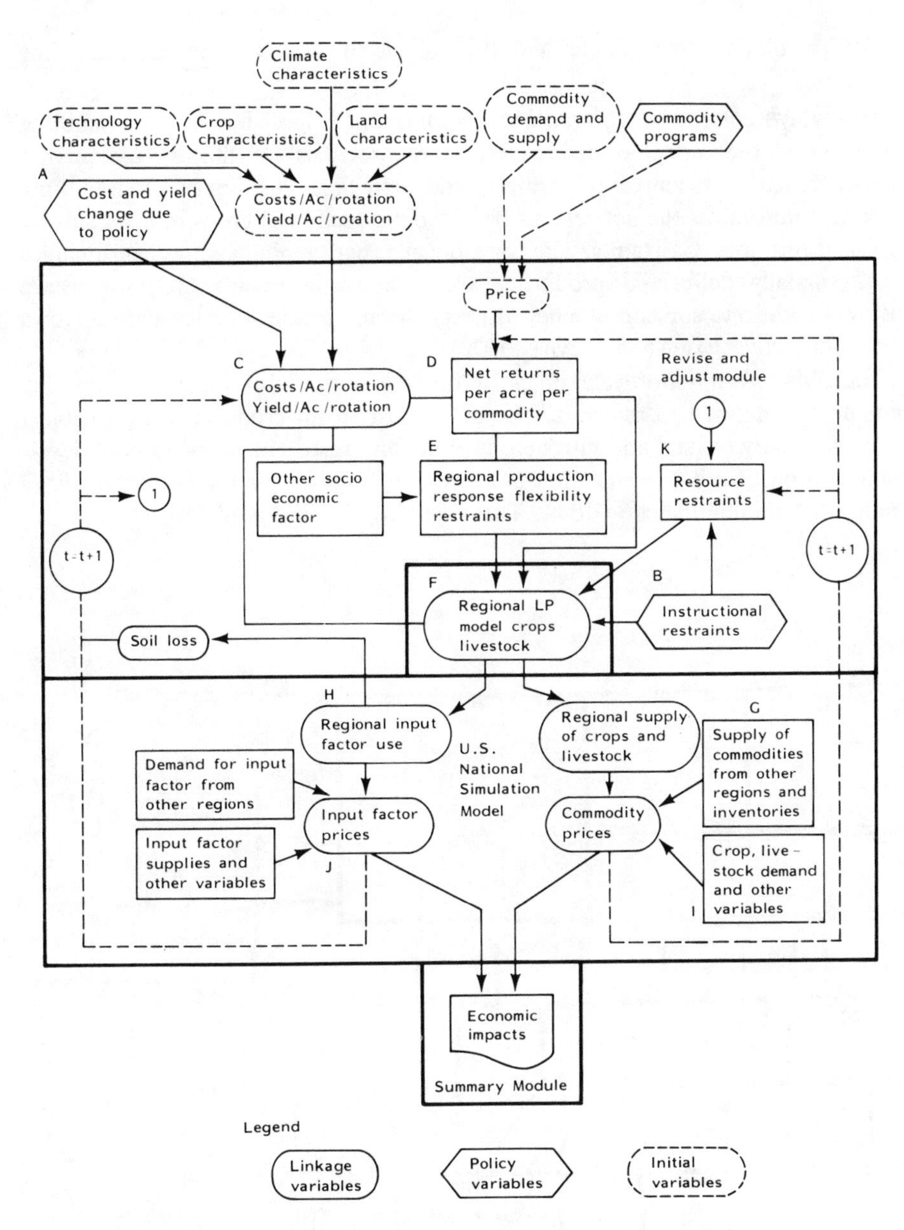

Figure 11.1. Structure of the regional-national recursive model.

Specification of the Model Components

Individual components of the model are discussed in detail in this section, beginning with the linear programming component. Attention is focused on the formulation of each component and the role it plays in the overall modelling framework.

The Iowa linear programming component. The linear programming component of the model is divided into three sections for this discussion: the objective function, resource constraints, and activities. The linear programming component maximizes the net returns from the production of crops in Iowa subject to a set of resource constraints. The Iowa linear programming component includes the 12 spatially delineated producing areas shown in Figure 11.2, which are consistent with Iowa soil conservancy districts enacted by the state legislature (Iowa State Cooperative Extension Service, 1972).

The objective function. The objective function of the Iowa linear programming component is defined to maximize the net returns from crop production subject to the availability of land and nitrogen fertilizer and restrictions placed upon levels of soil erosion. Profit maximization is a valid assumption for the commercialized farming that predominates in Iowa. The objective function is of the form:

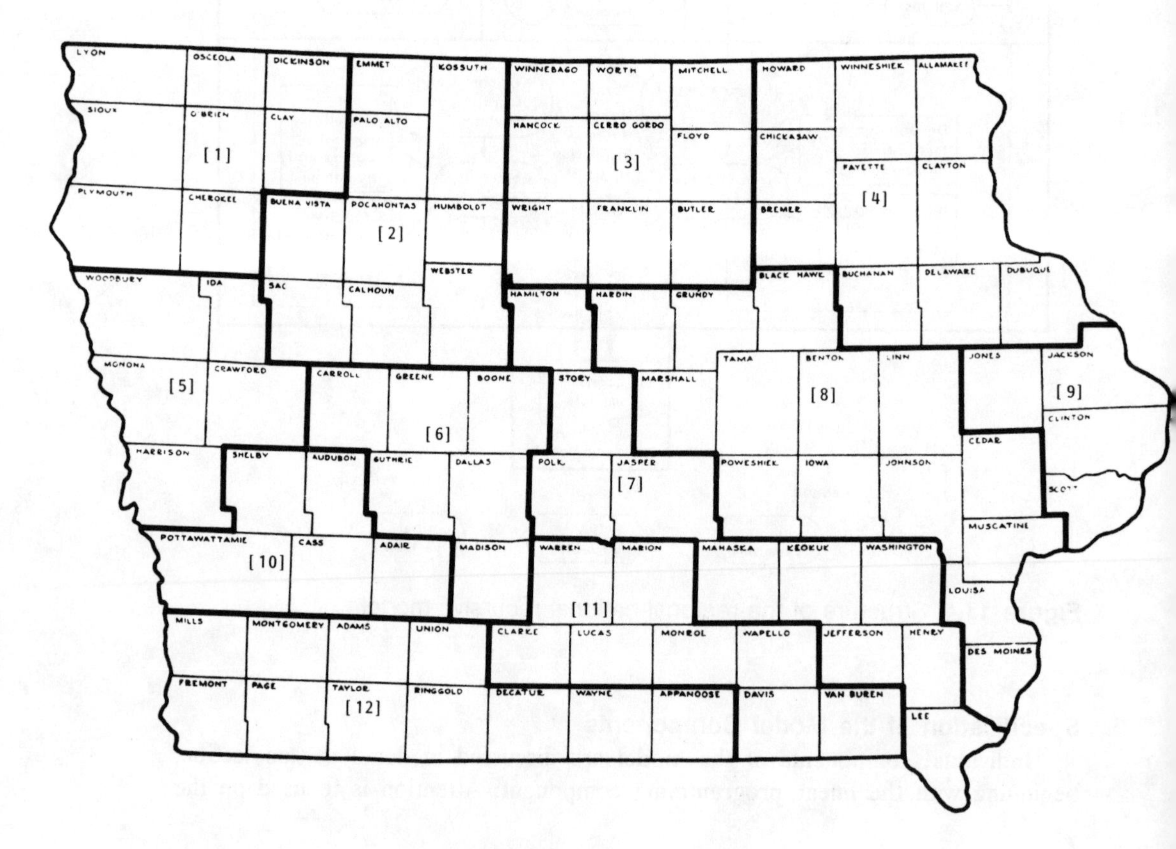

Figure 11.2. Iowa's producing areas.

$$\max OBJ = \sum_i \sum_j P^s_{ij} C^s_{ij} - \sum_i \sum_k \sum_l \sum_m T_{iklm} L_{iklm} - \sum_i P^n Q^{nb}_i \tag{11.1}$$

i = 1 to 12 for the producing areas;
j = 1 to 8 for the crops produced;
k = 1 to 9 for the conservation-tillage practices;
l = 1 to 48 for the crop rotations in each producing area;
m = 1 to 5 for the land groups.

where $P^s_{ij}C^s_{ij}$ is the gross return received by farmers for selling crop j C^s_j at price P^s_j in producing area i; $T_{iklm}L_{iklm}$ is the cost of production T, dollars per acre of rotation l with conservation-tillage practice k on land group m in producing area i, multiplied by the level of crop production activity L; and $P^n Q^{nb}_i$ is the price of nitrogen fertilizers P^n multiplied by the quantity of nitrogen purchased Q_{nb} in producing area i.

Commodity prices incorporated in the objective function are estimated within the U.S. econometric simulation component. Costs of producing agricultural commodities include charges for labor, machinery, pesticides, and fertilizers other than nitrogen. Cost coefficients for these items in the objective function, expressed in 1975 dollars, are derived from the Firm Enterprise Data System (FEDS) (U.S. Department of Agriculture, 1976b) and Iowa State Cooperative Extension Service budgets (1975).

The cost coefficient for each activity is adjusted for land group and conservation-tillage practice based on information from the U.S. Soil Conservation Service. A base of straight-row cropping is used for conservation practices, and adjustments are made in machinery and labor efficiency for contouring and strip cropping. Similarly, adjustments are made for the tillage practices where conventional tillage with no residue management serves as the base. Variations include conventional tillage with residue management and reduced tillage. Reduced tillage operations are further adjusted to reflect the tradeoff between tillage operations and the use of pesticides for weed control (Meister and Nicol, 1975).

Constraints. The land base represents the major constraint on the productive capacity of the system. Land in the Iowa linear programming component is used solely for dryland cropping activities defined for the endogenous crops and nonrotation pasture. Five land groups (Table 8.1) representing an aggregation of the twenty-nine class-subclasses in the *National Inventory of Soil and Conservation Needs, 1967* (USDA, Conservation Needs Inventory Committee, 1971) are specified for each of the 12 producing areas. These 1967 data are updated to 1975 using data developed by Paul Rosenberry (Private communication with Paul Rosenberry, USDA, Economic Research Service, Iowa State University).

The National Inventory of Soil and Conservation Needs places all soils in eight capability classes. The risks of soil damage or limitations in use become progressively more severe from Class I (few limitations) to Class VIII (no beneficial agricultural uses). (See Table 8.1.) Soils in the first four classes under good management are capable of producing commonly cultivated field crops and pasture plants. Soils in Classes V, VI, and VII may be suited for field crops under highly intensive management involving elaborate soil and water conservation practices. Soils in Class VIII do not return onsite benefits for inputs of management for crops without major reclamation (Klingebiel and Montgomery, 1966). Additionally, four land capability subclasses are defined according to the general kinds of limitations upon agricultural use. Subclass e is made up of soils where the susceptibility to water or wind erosion is the dominant problem or limitation in their use. Subclass w is made up of soils in areas with poor soil drainage, wetness, or high water table. Subclass s includes soils with shallow rooting zones, low moisture-holding capacity, or low fertility difficult to correct. Finally, Subclass c is made up of soils where temperature or lack of moisture is the only major hazard or limitation in their use. These land capability class-subclasses are aggregated into land groups to define the land resource base for the Iowa linear programming component of the model. This aggregation is shown in Table 11.1.

Table 11.1. Organization of land capability class-subclasses into the land groups defined for the Iowa model

Land Group	Land Capability and Class-Subclass[a]
1	I
2	II; IIIw, s,c,; IV w,s,c,; V
3	III e
4	IV e
5	VI; VII; VIII

[a]As included in the National Inventory of Soil and Conservation Needs, 1967 (Conservation Needs Inventory Committee, 1971). Land Capacity Classes are indicated by Roman numerals I-VIII; Subclasses by letters e (erosion hazard), w (wetness), s (rooting-zone limitations), and, c (climate). No general types of limitations or hazards are defined for Land Class I.

The other set of constraints applied at the producing area level are the crop transfer equations. Crop transfer equations simulate the marketplace for 8 commodities produced in the linear programming component: corn grain, corn silage, leguminous hay, nonleguminous hay, oats, sorghum grain, soybeans, and wheat. These transfer rows for each crop for each producing area take the form:

$$\sum_k \sum_l \sum_m Y_{ijklm} L_{ijklm} - C^s_{ij} = 0 \tag{11.2}$$

where Y_{ijklm} is the yield per acre of crop j in rotation l with conservation-tillage practice k on land group m in producing area i; L_{ijklm} is the activity level; and C^s_{ij} is the gross return per unit of crop j sold in producing area i.

The two statewide constraints in the model are nitrogen used for crop production and soil loss resulting from crop production. The nitrogen fertilizer transfer constraint acts as a marketplace for the supply and demand of commercial fertilizers used for the production of crops. The general form of the nitrogen constraint at the state level is:

$$Q_i^{nb} - \sum_i \sum_k \sum_l N_{iklm} L_{iklm} \geq 0 \qquad (11.3)$$

where Q_i^{nb} is the quantity of nitrogen fertilizers purchased for use in crop production in producing area i; N_{iklm} is the nitrogen requirement from commercial sources necessary for crop rotation 1 with conservation-tillage practice k on land group m in producing area i; and L_{iklm} is the crop production activity level.

The soil-loss row is an accounting row to determine the quantity of soil lost in the production of crops. The form of the state soil-loss equation is:

$$\sum_i \sum_l \sum_m S_{iklm} L_{iklm} \geq 0 \qquad (11.4)$$

where S_{iklm} is the quantity of gross soil loss occurring for crop activity i with conservation-tillage practice k on land group m in producing area i; in tons per acre; and L_{iklm} is the crop activity level.

Activities. Activities in the Iowa linear programming component are divided into crop production, crop marketing, and nitrogen purchase activities. Crop production activities simulate rotations producing corn grain, corn silage, legume and nonlegume hay, oats, sorghum grain, soybeans, and wheat in crop management systems incorporating rotations of one to four crops. Each rotation is defined for one of three conservation methods: straight-row, strip cropping, and contour plowing. Each conservation method can be defined with one or three tillage practices: conventional tillage, residue management, and reduced tillage. Appropriate combinations of conservation and tillage practices are defined on the land group to which they would apply. Thus each rotation combined with a specific conservation-tillage practice defines a unique crop management system:

$PARATIO_{in}$ = average fertilizer ratio applied for crop n in producing area i (pounds per acre);
$FERT_{in}$ = quantity of fertilizer applied in 1974 on crop n in producing area i (tons) (U.S. Department of Agriculture, 1976a);
$ACRE_{in}$ = number of acres on which fertilizer was applied on crop n in producing area i (acres) (U.S. Department of Commerce, 1977).

A weight is derived for each producing area and crop using equation 11.5:

$$TRATIO_{in} = PARATIO_{in}/SRATIO_n \qquad (11.5)$$

The 1974 state per-acre average for nitrogen, phosphorus, and potassium are found for each crop and weighted by the TRATIO's to give per acre averages for each producing area (English et al., 1980).

Crop yields are estimated using 1970-75 average county yields (Iowa Crop and Livestock Reporting Service, 1978-80). These county yields are weighted together using average county production to obtain producing-area yields for each crop. With the exception of hay, yields are adjusted for land group and conservation-tillage practice. Hay yields are derived from the *1974 Census of Agriculture* (U.S. Department of Commerce, 1977). The data obtained in a Soil Conservation Service questionnaire (Meister and Nicol, 1975) includes a set of ratios giving the relative land group yields of each crop category as compared to the most productive land group of the area. These ratios initially are weighted to the 5 land groups and are adjusted such that land group 1 has a relative yield value of 1.0. The conservation-tillage yield ratios are used equally on each land group to adjust yields of each crop (except hay) for both conservation and tillage effects. These ratios are scaled such that a representative "composite" conservation-tillage practice in each producing area equals 1.0.

Gross soil loss for the major land resource areas in Iowa is determined using the Universal Soil-Loss Equation (USLE). Soil-loss data are then weighted to the producing areas and incorporated in the appropriate crop management system. Calculated gross soil loss represents the average annual tons of soil leaving the field. Numerical values for each of the 6 factors in the USLE have been determined from research data (Wischmeier and Smith, 1965).

The variables in the USLE are defined as the dominant value existing on each soil class and subclass in each producing area. Soil loss is then computed by land resource area for each feasible combination of crop rotation, conservation-tillage practice, and soil class defined by the Soil Conservation Service (Meister and Nicol, 1975). The soil loss defined for the relevant twenty-nine soil class-subclasses is aggregated to get soil loss by the 5 land groups defined in the Iowa linear programming component. These coefficients are incorporated in the appropriate crop production activity and reflect the severity of erosion for the conditions on which the crop management system is specified.

The U.S. Econometric Simulation Component

The purpose of the U.S. econometric simulation component is to estimate resource use and commodity output originating in the United States excluding Iowa. These estimates are summed with the projections from the programming component for the state of Iowa to determine economic variables at the national level. The econometric component is based upon the CARD-National Agricultural Econometric Simulation Model (CARD-NAES) originally specified by Schatzer et al. (1981) and Roberts and Heady (1979, 1980), with some restructuring for this study. CARD-NAES predicts farmers' behavior in response to changing variables.

Agricultural production is included in the simulation system in 5 crop submodels (feed grains [corn, sorghum, oats, and barley], wheat, soybeans, cotton, and tobacco); 5 livestock submodels (beef, pork, lamb and mutton, chicken, and turkeys); and a submodel which aggregates variables from each of these 10 submodels with the exogenously determined variables for the rest of the U.S. agricultural sector. The submodels can be described in general terms as follows: resource demands in the current year depend directly or indirectly on lagged commodity and resource prices, lagged resource demands, and other variables;

current production depends upon the current quantity of resources demanded; supply in the current year depends on current production, carryover, and imports; average current-year commodity prices depend on current supply, exports, and other variables; commodity demand in the current year depends on current prices and other variables; gross income in the current year depends on current price and production; and quantity supplied is required to equal quantity demanded primarily through inventory adjustments.

Each crop submodel is divided into three stages corresponding to the pre-input (planning), input (planting), and output (harvesting and marketing) decisions in a one-year production cycle. From this comes the time path for each endogenous variable (i.e., variables whose values are determined within the model). The livestock submodels are aggregated into a single commodity group for the pre-input and input stages, but are disaggregated in the output stage.

The pre-input stage determines the level of physical assets committed to the production of farm commodities. Acreage intended for harvest, machinery purchases, machinery stocks, onfarm commodity stocks, value of land and buildings per harvested acre, average machinery and commodity stocks, total land and building value, and stocks of physical assets are determined in the pre-input stage for each endogenous crop and for aggregate livestock.

The levels of variable inputs needed to produce each endogenous crop are estimated in the input sector of each commodity submodel based on the level of fixed resources from the pre-input sector and other variables. The variable inputs considered include real estate taxes and expenditures, fertilizer, seed, fuel, oil, repairs, machinery, man-hours of labor, interest on commodity stocks, and miscellaneous inputs.

Estimates of prices received by farmers, commercial demand, supply, end-of-year stocks, and gross farm income resulting from resources committed in the pre-input and input stages are determined in the output sector of each commodity submodel. Livestock is also disaggregated into beef, pork, lamb and mutton, chicken, and turkey in the output stage of the livestock submodels.

Crop yields per harvested acre are incorporated into the output sector of each crop submodel using either of two approaches. In the first approach, crop yields are projected exogenously as linear functions of time using 1949-76 data. Then year-to-year deviations from those trend yields are determined from a production function that uses estimated elasticities of production for 6 inputs from the input sector. The production function used is:

$$Y_t = YB_t * 1.0 + \sum_{i=1}^{6} E_i * \left(\frac{I_{it} - B_{it}}{B_{it}}\right) \tag{11.6}$$

where Y_t is the yield per harvested acre; YB_t is the base run yield; E_i is the elasticity of production of the ith input (i = fertilizer, seed, labor, machinery, fuel, oil, repairs, and miscellaneous expenses); and B_{it} is the base run level of the ith input, all for each crop in year t. The input elasticities of production are estimated from factor share data using Tyner and Tweeten's methodology (1965).

In the second approach, crop yields are estimated endogenously as a function of the ratio of nitrogen price in year t (NP_t) and the commodity prices in period t - 1 (P_{t-1}), i.e., NP_t/P_{t-1}; a time trend variable to account for technological factors; and a proxy variable for weather (percentage deviation from normal pasture conditions). At the time of planting, the price of nitrogen NP_t is known to the farmer, while P_{t-1} is used as the expected price to be received for the crop when harvested. The ratio NP_t/P_{t-1} is assumed to be negatively related to yield in that when the nitrogen price increases relative to the expected commodity price, farmers tend to apply less nitrogen and hence receive a relatively lower yield.

CARD-NAES consists of 210 equations (151 for crops and 59 for livestock) formulated primarily in a sequential framework. Annual time series data are used to estimate the structural parameters of the model using appropriate statistical estimation techniques, including ordinary and autoregressive least squares and two-stage and three-stage least squares (Roberts and Heady, 1979). Most equations are estimated from 1949-76 data with portions of the livestock submodels using 1953-76 data.

The CARD-NAES model is adapted so that estimates of certain resource inputs and commodity output levels originating in the United States excluding Iowa can be obtained. In the current version of the model a distinction is made between Iowa and the rest of the United States only for harvested acreage, nitrogen fertilizer expense, and the quantity of each endogenous crop produced because these values are determined for the state of Iowa from the programming component. As the Iowa regional programming component is expanded in future versions, other categories of resource inputs and commodity outputs can be differentiated within the regional-national system.

Acreage of feed grains, wheat, and soybeans intended for harvest for the United States excluding Iowa is determined by historical shares based upon the following regression equation:

$$AC47_i = \beta_i * AC48_i \qquad (11.7)$$

where $AC47_i$ is the estimated harvested acreage of crop i (i = feed grains, wheat, soybeans) for the United States excluding Iowa; $AC48_i$ is the estimated harvested acreage for the United States including Iowa; and β_i is the estimated share of total U.S. acreage for crop i not harvested in Iowa, $0 < \beta_i \leq 1$. No cotton or tobacco is grown in Iowa, hence $\beta_i = 1.0$ for these crops. The national harvested acreage for each endogenous crop is then determined by equation (11.8), where $AC47_i$ is found from equation (11.7) and:

$$AC48_i = AC47_i + AC01_i \qquad (11.8)$$

$AC01_i$ is the acreage of crop i harvested in Iowa (derived by the summation of production activity levels for each crop in the regional programming component).

Nitrogen fertilizer expense for the United States excluding Iowa is found by the product of the estimated fertilizer expense per acre and the summation of the acreage estimates for each crop for the United States less Iowa (derived in the pre-input state). Aggregate nitrogen purchases in Iowa are found within the programming component by a summation of nitrogen buying activities for each producing area, and then added with the econometric estimates for the rest of the United States to obtain a national fertilizer expense level.

Production or supply of each crop for the United States excluding Iowa is estimated through the econometric model by multiplying the harvested acreage of each crop by the estimated yield. Crop production for Iowa is found by summing activity levels for each crop in the programming component.

Changes in input factor costs and commodity prices are found within the econometric component for subsequent use in revising the programming component. A discussion of this procedure is presented in relation to the linkage component.

The Linkage Component

The purpose of the linkage component of the Iowa regional-national system is to transfer information between the programming and econometric components and to revise and adjust selected variables between time periods to simulate the recursive sequence of agricultural production and its interaction with the environment.

The basic solution procedure for the model is shown in Figure 11.3. The regional linear programming component is first solved for the profit-maximizing level of crop production and resource use for the state of Iowa. These values are summed with estimates of production and input use occurring in the United States excluding Iowa (estimated from the national econometric simulation component) to obtain national totals. Commodity prices and other important economic variables are estimated in the econometric component. Crop yield adjustment factors are determined in the physical component based on inches of top soil lost and are used to revise the crop yields in the linear programming sector. The newly estimated commodity prices are used to revise the crop-marketing coefficients in the objective function of the Iowa programming component. After the linear programming input data matrix has been revised, the programming component is solved for the next time period, thus repeating the entire process until the predetermined number of simulations are completed.

The linkage component can be separated into three subsectors: retrieval, adjustment, and revision. Linkage variables, defined as variables providing information from the regional model to the national model and vice versa, are specified for these three subsectors. In the current version of the Iowa model, information retrieved from the Iowa programming component includes production levels of endogenous crops, soil loss, nitrogen fertilizer use, and land use in each of 5 land groups for each of 12 producing areas. Crop production and fertilizer use are inputs to the econometric component, while soil loss and land use are inputs to the adjustment and revision subsectors of the linkage component.

The purpose of the adjustment subsector is to adjust the estimated crop yields

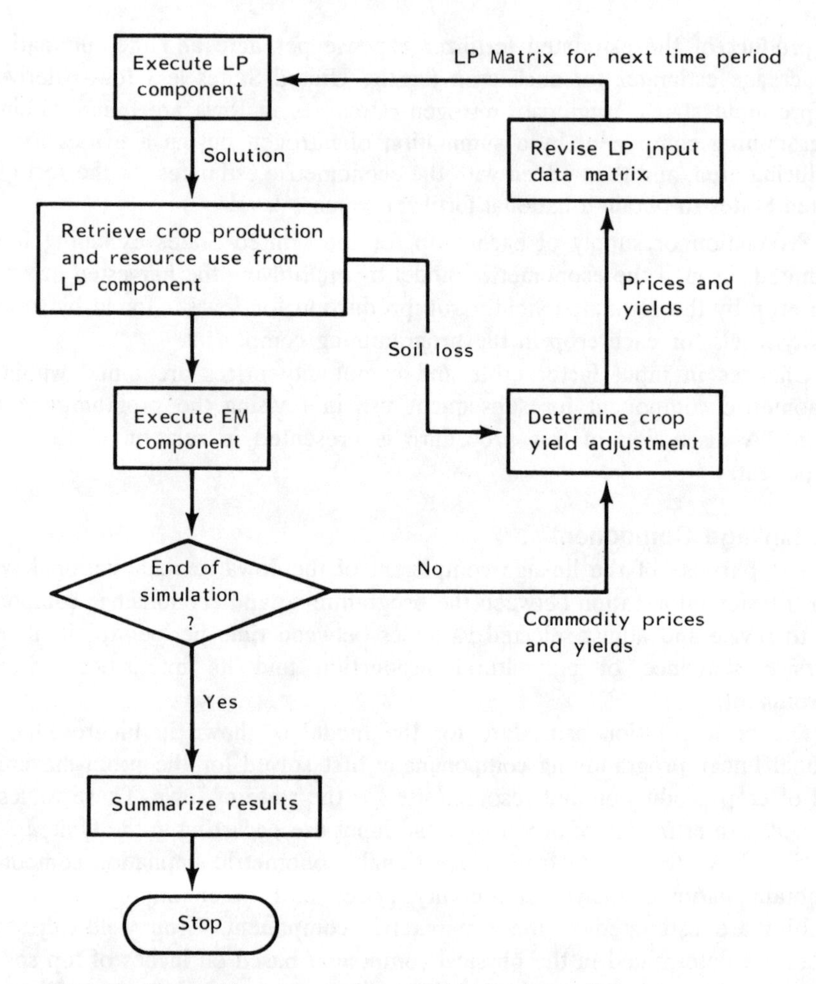

Figure 11.3. Basic solution procedure of the Iowa model.

to account for the effects of soil loss. The hypothesized relationship between topsoil depth and the yield of corn, for example, is presented in Figure 11.4. Topsoil depths of 1.5, 5.0, and 9.5 inches correspond to average soil depths of erosion phases 3 (severely eroded), 2 (moderately eroded), and 1 (slightly eroded), respectively. This study assumes that changes in topsoil depth above 9.5 inches or below 1.5 inches do not affect crop yields. Available empirical evidence tends to support this assumption. Benchmark corn yield-soil depth functions similar to Figure 11.4 are estimated for each of 5 land classes for each of 12 producing areas based on the principal soil association areas in each producing area, the dominant soil classification by land class for each land resource area, and information on 18 Iowa farms from Pope, Bhide, and Heady (1982).

Initial yields in the linear programming model are based on the 1970-75

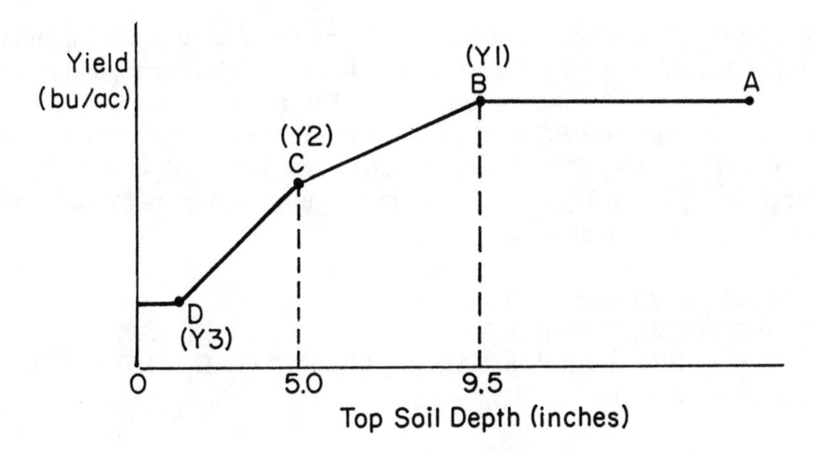

Figure 11.4. Hypothesized relationship between topsoil depth and corn yields.

average for the 8 crops in the 12 producing areas, adjusted for land-class differentials. These initial yields are then adjusted for each crop between time periods based on two factors. First, average yields per acre per land class per producing area increase over time because of technological improvements in crop varieties, etc. This improvement essentially amounts to a gradual horizontal upward shift of curve DCBA in Figure 11.4. Hence, technological improvement is expected to allow more bushels per acre of the respective crop to be produced from a given topsoil depth.

A second adjustment procedure allows decreases in crop yields over time due to loss of topsoil. The procedure used to make these adjustments is as follows. The soil loss estimated by the Universal Soil-Loss Equation in tons per acre is converted to inches of soil loss per acre using equation (11.9):

$$ISL_{im} = \frac{TSL_{im}}{SBD_{im} * TACRE_{im}} \tag{11.9}$$

where ISL_{im} is the estimated inches of soil lost in producing area i on land group m; TSL_{im} is the total tons of soil lost; SBD_{im} is the average bulk density of the soil; and $TACRE_{im}$ is the total acres of cropland.

Once inches of soil loss is determined from equation (11.9), the revised soil depth is determined by equation (11.10):

$$SOILD_{imt} = SOILD_{imt-1} - ISL_{imt} \tag{11.10}$$

where $SOILD_{imt}$ is the average soil depth of land group m in producing area i at time t. Let Y1, Y2, and Y3 in Figure 11.4 be the corn yields corresponding to soil depths of 9.5, 5.0, and 1.5 inches, respectively. The slope over line segment AB = M(1) = 0; the slope over BC = M(2) = ((Y1 - Y2)/Y1)/4.5; the slope over CD = M(3) = ((Y2 - Y3)/Y1)/3.5; and the slope over DE = M(4) = 0. Based on the benchmark plots in Figure 11.4, the yield adjustment factors ($YADJ_t$) due to loss of topsoil are computed as:

If $SOILD_t \geq 9.5$ then $YADJ_t = 1.0$;
If $5.0 \leq SOILD_t \leq 9.5$ then $YADJ_t = M(2) * (SOILD_t - 5.0) + (Y2/Y1)$;
If $1.5 \leq SOILD_t \leq 5.0$ then $YADJ_t = M(3) * (SOILD_t - 1.5) + (Y3/Y1)$;
If $SOILD_t < 1.5$ then $YADJ_t = 1.0$.

Using the appropriate yield adjustment factor determined above on the basis of topsoil depth ($SOILD_t$), crop yields are determined for each crop in each land class in each producing area as:

$$CYLD = IYLD * (1 - YADJ);$$
$$YIELD_t = IYLD' - (CYLD * WGT)$$

where CYLD is the estimated change in crop yield; IYLD is the initial yield based on the 1970-75 average crop yields in Iowa; IYLD' is the initial yield adjusted for conservation-tillage practice; and WGT is the weight of the crop in the particular rotation. To summarize, the primary purpose of the yield adjustment subsector of the Iowa regional national recursive model is to determine the net adjustment in crop yields between time periods. The net adjustment is the difference between the technological improvement and the soil-loss detriment.

The revised subsector of the linkage component takes information from the retrieval and the yield adjustment subsectors and revises the linear programming component for the next time period. The revision reflects the expected yield and commodity prices determining the next period's production response.

Results of Alternative Test Solutions

The Iowa regional-national model is used to investigate crop production activity in the state of Iowa under two alternative test solutions. The focus is on studying the economic impact of alternative policies directed at increasing the long-run sustainability of agricultural production through erosion control. Alternative I is a baseline run with no restrictions on soil loss. This solution indicates the profit-maximizing level of production of the 8 endogenous crops and the resulting soil loss for each producing area in Iowa when no restrictions are placed on the amount of soil leaving the field via water erosion. Alternative II limits soil loss by prohibiting the use of any crop management system that is estimated to result in a soil loss of greater than 5 tons per acre. The solution to this alternative indicates the changes in crop production in each producing area (i.e., changes from Alternative I) that are necessary to meet the 5 ton per acre criteria.

A test run for the years 1980-2000 is conducted and results obtained for each of 12 producing areas in Iowa. In these alternatives the econometric model is solved every year to determine crop production levels for the United States excluding Iowa, livestock production for the United States, commodity prices, and other variables. The Iowa regional programming model is solved once every 5 years to determine Iowa crop production of corn, oats, sorghum, soybeans, wheat, and legume and nonlegume hay; total acres used for crop production; acres left idle (available for crop production but not used); total soil loss; average soil loss per acre; and quantities of commercial nitrogen purchased. Due to space limitations, only selected results are presented here. More details are available from the authors.

Estimated crop production in Iowa for Alternatives I and II for the years 1980 and 1985 are presented in Table 11.2. Removing cropping practices that lead to soil losses greater than 5 tons per acre (Alternative II) noticeably affects oats, soybean, and wheat production. Soil-saving production practices employed in Alternative II typically include small grains in various rotations with row crops. Corn maintains its comparative advantage in most producing areas and is produced primarily with reduced-tillage cropping practices. The most significant changes in cropping practices occur, as expected, in producing areas experiencing the highest level of soil erosion, producing areas 5 and 10 in western Iowa.

Cropland acreage, soil loss, and commercial nitrogen purchases for Iowa in 1980 and 1985 under Alternatives I and II appear in Table 11.3. The estimates of average soil loss per acre in Alternative I (with no erosion restrictions) range from an average of 2.86 tons per acre in producing area 2 to 14.08 tons per acre in producing area 10. Eight of the 12 producing areas have losses exceeding the 5 tons per acre goal set by the Iowa State Legislature.

Prohibiting the use of crop management practices that result in losses in excess of 5 tons per acre in Alternative II greatly reduces the level of soil loss in each producing area in Iowa. For example, in producing area 10 soil loss drops from an average of 14.08 tons per acre to 3.23 tons per acre. This reduction is due both to erosive land going out of production and changes in crop management practices. Limiting crop production to management practices having an average soil loss of less than 5 tons per acre reduces the average soil loss for each producing area well below the 5-ton limit.

Some changes in the level of commercial nitrogen fertilizer purchases for each producing area occur between Alternatives I and II. These changes correspond to increases or decreases in corn production in the respective producing area.

Soil erosion is dependent upon the topography of the cropland, the crops produced, and the management and tillage practices employed. Disaggregation of average soil loss per acre to the producing area and land-class level indicates the diversification of soil erosion across the state of Iowa (Table 11.4). As the

Table 11.2. Crop production in Iowa for Alternative I (Alt. I) and Alternative II (Alt. II) fo 1980 and 1985 by producing area

PA	Year	Corn		Oats		Soybeans		Wheat	
		Alt. I	Alt. II	Alt. I	Alt. II	Alt. I	Alt. II	Alt. I	Alt. II
		(thousand bushels)							
1	1980	194,585	182,375	1,200	5,715	19,686	16,462	0	0
	1985	165,358	153,149	1,200	5,715	29,452	26,228	0	0
2	1980	218,099	214,892	1,781	1,781	981	1,892	1,039	1,039
	1985	103,621	104,945	1,781	1,764	33,943	33,491	1,041	1,076
3	1980	284,631	283,278	1,351	1,351	1,216	1,605	1,373	1,372
	1985	136,666	137,673	1,351	1,338	43,366	43,033	1,373	1,410
4	1980	148,541	159,199	874	874	22,782	19,567	6,789	6,783
	1985	82,644	90,652	874	1,125	42,663	39,689	6,789	6,862
5	1980	73,208	63,262	515	719	14,313	6,227	0	0
	1985	73,208	46,583	515	719	14,313	9,377	0	2,816
6	1980	60,408	67,346	2,220	2,199	24,951	19,576	1,990	2,025
	1985	58,121	66,070	2,220	2,199	25,671	21,553	1,990	3,114
7	1980	90,923	121,427	2,179	2,153	42,558	31,397	4,397	4,504
	1985	90,923	99,143	2,179	2,160	42,558	34,828	4,397	9,020
8	1980	149,845	220,398	4,137	4,098	65,010	40,203	5,048	5,426
	1985	149,845	144,303	4,137	5,303	65,010	49,000	5,048	15,660
9	1980	30,747	49,279	1,570	2,088	15,511	8,848	1,668	1,559
	1985	30,747	38,278	1,570	1,596	15,512	11,050	1,668	3,058
10	1980	63,820	54,564	816	1,142	11,676	4,685	0	0
	1985	63,820	39,231	816	1,142	11,676	7,385	0	2,582
11	1980	9,530	18,244	2,664	3,157	11,751	9,110	5,148	4,741
	1985	4,421	17,563	2,663	2,639	13,290	9,281	5,148	5,124
12	1980	81,531	66,435	0	2,221	13,815	11,698	3,393	4,872
	1985	82,485	66,435	2,679	2,221	14,362	11,698	26	4,872
TOTAL	1980	1,405,686	1,509,699	19,307	27,798	244,250	171,270	30,845	32,321
	1985	1,041,859	1,044,025	21,985	27,921	351,816	296,583	27,480	55,594

profitability of alternative crops changes over time, so do production levels and the resulting soil loss. Limiting soil loss to 5 tons per acre in Alternative II increases soil loss on some land classes in some producing areas. This increase is a result of crops shifting away from land classes 3 and 4 toward the higher yielding land in classes 1 and 2. Land class 5 generally has such low crop yields that it is not profitable for crop production.

This analysis was made for Iowa as a region within a national framework to conform with a project that the International Institute of Applied Systems Analysis (IIASA) has promoted among a group of studies. The purpose of the study is one of determining sustainable production systems. We could, for example, raise able 11.3. Cropland acreage, soil loss, and nitrogen purchased in Iowa for 1980 and 1985 by producing area

		Acreage Planted to Crops		Cropland Not Used		Total Soil Loss		Average Soil Loss/Acre		Nitrogen Purchased	
	Year	Alt. I	Alt. II	Alt. I	Alt. II	Alt. I	Alt. II	Alt. I	Alt. II	Alt.I	Alt.II
		(thousand acres)				(thousand tons)		(tons/acre)		(thousand lbs.)	
	1980	2,382	2,350	9	41	16,420	6,667	6.89	2.84	207	194
	1985	2,382	2,350	9	41	16,328	7,575	7.27	3.22	180	167
2	1980	1,957	1,957	12	12	3,919	3,951	2.00	2.02	288	284
	1985	1,957	1,957	12	12	7,286	5,939	3.72	3.03	144	164
3	1980	2,386	2,386	9	9	5,552	5,232	2.33	2.19	384	382
	1985	2,386	2,386	9	9	9,555	8,375	4.00	3.51	191	193
4	1980	2,347	2,347	68	68	9,324	7,056	3.97	3.01	212	226
	1985	2,347	2,347	68	68	12,673	8,624	5.40	3.67	133	145
5	1980	1,292	938	168	522	17,186	2,851	13.30	3.04	105	85
	1985	1,292	938	168	522	17,186	2,637	13.30	2.81	105	68
6	1980	1,347	1,347	16	16	7,525	4,206	5.59	3.12	78	97
	1985	1,347	1,347	16	16	7,728	4,530	5.74	3.36	76	88
7	1980	2,269	2,269	60	60	14,678	9,011	6.47	3.97	141	185
	1985	2,269	2,269	60	60	14,678	9,641	6.47	4.25	141	161
8	1980	3,288	3,288	121	121	20,019	11,172	6.09	3.40	206	294
	1985	3,288	3,288	121	121	20,019	11,933	6.09	3.63	206	213
9	1980	850	850	61	61	5,702	2,728	6.71	3.21	44	64
	1985	850	850	61	61	5,702	2,912	6.71	3.43	44	567
0	1980	1,197	881	163	478	16,848	2,946	14.08	3.34	87	69
	1985	1,197	881	163	478	16,848	2,744	14.08	3.11	87	56
1	1980	981	981	81	81	3,234	1,964	3.30	2.00	26	37
	1985	981	981	81	81	3,553	1,950	3.62	1.99	20	36
2	1980	1,738	1,571	65	231	15,803	5,067	9.09	3.23	110	90
	1985	1,738	1,571	65	231	8,297	5,067	4.77	3.23	106	90
otal	1980	22,034	21,165	833	1,700	136,210	62,851	6.18	2.97	1,888	2,007
	1985	22,034	21,165	833	1,700	140,853	71,927	6.39	3.40	1,433	1,419

production to high levels in the near future only to have production in the longer run decline due to soil erosion, soil salinity, desertication, and other reasons. CARD at Iowa State University already has the capacity to make such a study on a national basis with as many as 233 producing areas with 12 land groups in each, many market and water supply regions, and a vast set of endogenous commodities. For an illustration of how such combined programming and econometric models might be applied on a national and interregional basis, see Langley et al. (1982), Huang and Heady (1982), and Heady et al. (1983).

Work is currently in progress to expand and improve the existing structure of the model. Work on the U.S. national simulation sector is primarily directed toward improving the estimation of acreage response functions for the continental

Table 11.4. Soil loss per acre over time in Iowa by producing area and land class

PA	Land class	1980 Alt. I	1980 Alt. II	1985 Alt. I	1985 Alt. II	1990 Alt. I	1990 Alt. II	1995 Alt. I	1995 Alt. II	2000 Alt. I	2000 Alt. II
		(tons per acre)									
1	1	4.35	4.66	4.66	4.66	4.55	4.66	4.66	4.66	4.66	3.28
	2	1.68	2.19	1.95	2.05	1.95	2.22	1.95	1.95	1.95	1.95
	3	32.64	2.52	25.48	3.92	27.67	1.69	27.61	2.24	28.80	3.61
	4	--	--	7.35	--	7.35	--	7.35	--	7.35	4.90
	5	--	--	--	--	--	--	--	--	3.79	--
2	1	0.36	4.13	0.36	4.13	0.36	4.13	0.36	4.13	0.36	2.46
	2	3.12	3.04	3.11	3.04	3.11	3.04	3.11	3.04	3.11	1.73
	3	6.64	2.63	7.03	2.79	7.39	2.96	7.71	3.12	7.98	3.33
	4	19.01	2.85	19.01	2.85	19.01	2.85	19.01	2.85	19.01	1.90
	5	2.98	2.98	2.98	2.98	2.98	2.98	2.98	2.98	2.98	2.98
3	1	4.09	4.44	4.15	4.50	4.20	4.50	4.24	4.50	4.31	2.51
	2	2.94	2.99	2.94	2.99	2.94	2.99	2.94	2.99	2.94	1.70
	3	6.51	1.96	6.86	2.23	6.97	2.44	7.05	2.63	7.12	2.98
	4	--	--	17.62	2.64	17.62	2.64	17.62	2.64	17.62	1.76
	5	--	--	--	--	--	--	--	--	4.20	4.20
4	1	5.75	2.23	6.08	4.07	3.97	4.07	4.09	4.07	4.82	2.98
	2	3.23	3.35	3.23	3.35	3.57	3.35	3.57	3.35	3.57	1.85
	3	5.76	4.21	6.25	4.05	6.40	4.21	6.51	4.21	6.61	4.00
	4	--	--	17.55	2.63	17.55	2.63	17.55	2.83	17.55	2.14
	5	--	--	--	--	--	--	--	--	--	--
5	1	6.15	3.42	6.20	4.89	6.61	3.91	6.61	3.91	6.61	4.65
	2	1.45	2.15	1.97	2.15	1.97	2.15	1.97	2.15	1.97	1.38
	3	37.88	4.41	30.09	1.83	30.69	1.82	31.19	1.81	31.61	3.96
	4	--	--	--	--	35.02	--	35.91	--	36.67	--
	5	--	--	--	--	--	--	--	--	--	--
6	1	3.60	3.20	3.83	3.99	4.02	3.87	5.97	4.00	5.97	3.04
	2	3.55	3.70	3.55	3.69	3.55	3.69	3.55	3.69	3.55	2.30
	3	8.80	2.00	8.96	1.40	9.11	2.52	3.95	2.01	4.49	2.26
	4	21.03	3.15	21.03	3.15	21.03	3.15	21.03	3.15	21.03	3.15
	5	--	--	--	--	--	--	6.19	--	6.19	--
7	1	4.48	4.56	6.84	4.69	6.84	4.76	6.84	4.76	6.84	3.63
	2	4.86	4.86	4.86	4.86	4.86	4.86	4.86	4.86	4.86	2.71
	3	9.42	3.55	5.54	2.75	6.06	3.11	7.10	3.33	7.49	3.41
	4	12.25	1.83	12.25	3.06	12.25	4.90	12.25	4.90	12.25	4.90
	5	--	--	--	--	--	--	5.58	--	5.58	--
8	1	6.69	3.64	6.69	4.49	6.69	4.58	6.68	4.58	6.68	3.49
	2	5.08	3.18	5.08	3.74	5.08	3.69	5.08	3.74	5.08	2.84
	3	3.97	3.07	4.45	2.51	4.93	3.15	6.11	3.07	6.50	3.11
	4	9.52	1.42	9.52	1.42	9.52	1.42	9.52	1.42	9.52	1.42
	5	--	--	--	--	--	--	6.26	--	6.26	--
9	1	5.02	2.23	0.75	2.34	1.47	2.25	1.66	2.29	3.56	4.77
	2	2.09	2.52	4.89	4.29	4.80	4.29	4.89	4.29	4.89	4.39
	3	5.32	3.98	8.56	3.89	8.78	4.02	8.86	4.19	8.93	2.61
	4	11.79	--	11.79	1.76	11.79	4.71	11.79	4.71	11.79	1.76
	5	--	--	--	--	--	--	--	--	--	--

Table 11.4. Continued

PA	Land class	1980 Alt. I	1980 Alt. II	1985 Alt. I	1985 Alt. II	1990 Alt. I	1990 Alt. II	1995 Alt. I	1995 Alt. II	2000 Alt. I	2000 Alt. II
						(tons per acre)					
10	1	5.99	3.63	7.01	4.93	7.01	4.15	7.01	4.15	7.01	4.93
	2	2.09	2.33	2.09	2.09	2.09	2.09	2.09	2.09	2.09	2.09
	3	30.67	3.60	32.69	3.38	33.68	3.08	32.84	3.08	34.76	4.19
	4	--	--	24.90	--	27.21	--	34.08	--	35.66	--
	5	--	--	--	--	--	--	--	--	--	--
11	1	0.55	4.32	1.90	2.07	3.40	2.81	6.08	3.27	6.76	4.29
	2	5.52	3.32	5.42	3.78	5.18	3.78	5.14	3.78	5.14	2.87
	3	8.34	0.94	8.81	1.89	9.23	1.70	9.55	2.50	9.81	2.45
	4	--	1.97	7.99	3.19	7.99	3.19	7.99	4.38	7.99	4.50
	5	--	--	--	--	--	--	6.37	--	7.71	--
12	1	5.67	3.88	5.17	4.44	7.49	4.44	7.49	4.44	7.49	4.44
	2	3.11	3.12	3.11	3.11	3.11	3.11	3.11	3.11	3.11	2.18
	3	19.99	3.96	24.99	3.69	26.99	3.28	28.55	2.87	24.65	4.72
	4	--	--	37.84	--	8.11	--	12.13	4.50	37.84	4.50
	5	--	--	6.37	--	6.37	--	6.37	--	6.37	--

United States excluding Iowa, interactions of government policy variables with agricultural production, and the demand for input factors. The Iowa projection of the linear programming model, with Iowa taken as a single region, emphasizes crop and livestock production in relation to soil erosion or conservation and land productivity. The historical share approach to obtaining acreage estimates for the forty-seven states implicitly assumes that the national model dominates the projection process, while an implicit assumption of the regional-national structure is that the regional model should dominate for the production sustainability aspects. Hence, it is consistent with the model framework that the acreage response for the forty-seven states other than Iowa are estimated by the national econometric model. The Iowa programming responses then are added to these. Regional acreage response functions are estimated for each endogenous crop grown in Iowa.

The demand and supply of factors of production play an important role in the potential sustainability of agricultural production. In regards to soil loss, investment in terracing and other forms of land improvement will need attention. Olson (1980) offers suggestions from which to develop a means of estimating investment behavior for the purpose.

The estimates of soil loss per acre are currently based on the best judgment of experts working in this area. Procedures to improve these estimates are in progress. Additional information on the proposed soil-loss simulator may be found in Huang and rosenberry (1982).

The model specified in this chapter may be applied to the analysis of alternative scenarios concerning the interrelationships between agricultural production and the surrounding environment. Scenarios being analyzed include the following: restrictions on selected outputs from the production process, such as alternative soil loss limits (an example of which is presented in this paper); restrictions on selected inputs into the production process, e.g., the quantity of nitrogen fertilizer available for application; the impact of changes in relative input and output prices on production patterns in Iowa; subsidies and cost sharing for conservation practice; and taxation of soil loss, which reduces soil productivity (Kapar, 1983). Changes in certain exogenous variables and policy instruments in the econometric component will have various impacts upon relative prices paid for inputs and received for commodities and subsequently upon agricultural production practices in Iowa. After such changes, the impacts on soil loss, production levels, patterns, costs, and commodity prices are estimated.

REFERENCES

English, Burton C., Jeff Guernsey, and Earl O. Heady. 1980. Impact of Sheet and Rill Erosion on Agricultural Lands. Unpublished working paper. Center for Agricultural and Rural Development, Iowa State University, Ames.

Heady, Earl O., and James A. Langley. 1981. Specification of a Regional-National Recursive Model for IIASA/FAP's Iowa Task 2 Case Study. WP-81-90. International Institute for Applied Systems Analysis, Laxenburg, Austria.

Heady, Earl O., James A. Langley, and Wen-yuan Huang. 1981. A Research Adaptive Model for National and Interregional Analysis. In *Economics and Mathematical Systems*, edited by Josef Gruber. Springer-Verleg, Berlin.

Huang, Wen-yuan, and Earl O. Heady. 1980. Linkage of a National Recursive Econometric Model and an Interregional Programming Model for Agricultural Policy Analysis. In *Large-Scale Energy Models*, AAAS Selected Symposium 73, San Francisco. Westover Press, Boulder, Colorado.

Huang, Wen-yuan, and Paul Rosenberry. 1982. The Soil Loss Simulator. Unpublished paper. USDA, Natural Resource Economics Division and Iowa State Division, Iowa State University.

Iowa Crop and Livestock Reporting Service. 1978-80. Iowa Agricultural Statistics. Iowa Department of Agriculture, Des Moines.

Iowa State Cooperative Extension Service. 1975. Estimated costs of Crop Production in North Central Iowa. Iowa State University, Ames.

Johnston, J. 1972. *Econometric Methods*. 2nd Edition. McGraw-Hill, New York.

Kapur, Aasha. 1982. Application of Sustainable Agricultural Production Policies by Means of a Recursive Econometric Programming Model. Unpublished Ph.D. dissertation, Iowa State University, Ames.

Klingebiel, A. A., and P. H. Montgomery. 1966. Land-Capability Classification. USDA Agriculture Handbook No. 210. Washington, D.C.

Langley, James A., Burton C. English, and Earl O. Heady. 1982. The Impact of Alternative Water Policies on Agricultural Production in Nebraska: A Hybrid Model Approach. CARD Series Paper 82-10. Center for Agricultural and Rural Development, Iowa State University, Ames.

Langley, James A., Wen-yuan Huang, and Earl O. Heady. 1981. A Recursive Adaptive Hybrid Model for the Analysis of the National and Interregional Impacts of Three Alternative Agricultural Situations for 1981-83. CARD Report 100. Center for Agricultural and Rural Development, Iowa State University, Ames.

Meister, Anton D., and Kenneth J. Nicol. 1975. A Documentation of the National Water Assessment Model of Regional Agricultural Production, Land and Water Use, and Environmental Interaction. CARD Miscellaneous Report. Center for Agricultural and Rural Development, Iowa State University, Ames.

Olson, Kent D. 1980. The Resource Structure of United States Agriculture: An Economic Analysis. Unpublished Ph.D. dissertation, Iowa State University, Ames.

Roberts, Roland K., and Earl O. Heady. 1980. An Analysis of Selected Agricultural Policy on the U.S. Livestock Sector by an Econometric Simulation Model. CARD Report 92. Center for Agricultural and Rural Development, Iowa State University, Ames.

Schatzer, Raymond Joe, Roland K. Roberts, Earl O. Heady, and Kisan R. Gunjal. 1980. An Econometric Simulation Model to Estimate Input Stocks and Expenses, Supply Response, and Resource Demand for Several U.S. Agricultural Commodities. CARD Report 102. Center for Agricultural and Rural Development, Iowa State University, Ames.

Tyner, Fred C., and Luther G. Tweeten. 1965. A Methodology for Estimating Production Parameters. *Journal of Farm Economics* 47:1462-67.

U.S. Department of Agriculture, Conservation Needs Inventory Committee. 1971. National Inventory of Soil and Conservation Needs, 1967. Statistical Bulletin No. 461. Washington, D.C.

U.S. Department of Agriculture, Economic Research Service. 1976a. Energy and U.S. Agriculture, 1974. Data base. Washington, D.C.

___. 1976b. Firm Enterprise Data System. Oklahoma State University, Stillwater.

U.S. Department of Commerce, Bureau of the Census. 1977. *1974 Census of Agriculture.* Volume 1: State and County Data. Part 15, Iowa. U.S. Government Printing Office. Washington, D.C.

Wischmeier, W. H., and D. D. Smith. 1965. Predicting Rainfall Erosion Losses from Cropland East of the Rocky Mountains. USDA Agriculture Handbook No. 282. Washington, D.C.

INDEX